Recent Advances in 3D Printing Technologies in Bioengineering with Selected Papers from the 29th Congress of the European Society of Biomechanics (ESB 2024)

Recent Advances in 3D Printing Technologies in Bioengineering with Selected Papers from the 29th Congress of the European Society of Biomechanics (ESB 2024)

Guest Editors

Bin Zhang
Amirpasha Moetazedian
Peter Zioupos

MDPI

Basel • Beijing • Wuhan • Barcelona • Belgrade • Novi Sad • Cluj • Manchester

Guest Editors

Bin Zhang
Department of Mechanical and Aerospace Engineering
Brunel University London
London
UK

Amirpasha Moetazedian
School of Engineering
University of Hull
Kingston Upon Hull
UK

Peter Zioupos
School of Engineering
University of Hull
Kingston Upon Hull
UK

Editorial Office
MDPI AG
Grosspeteranlage 5
4052 Basel, Switzerland

This is a reprint of the Special Issue, published open access by the journal *Journal of Manufacturing and Materials Processing* (ISSN 2504-4494), freely accessible at: https://www.mdpi.com/journal/jmmp/special_issues/7F2RXWA2WA.

For citation purposes, cite each article independently as indicated on the article page online and as indicated below:

Lastname, A.A.; Lastname, B.B. Article Title. *Journal Name* **Year**, *Volume Number*, Page Range.

ISBN 978-3-7258-7238-1 (Hbk)
ISBN 978-3-7258-7239-8 (PDF)
https://doi.org/10.3390/books978-3-7258-7239-8

Cover image courtesy of Peter Zioupos

Contents

About the Editors

Bin Zhang

Bin Zhang is an assistant professor in Additive Manufacturing at the Department of Mechanical and Aerospace Engineering at Brunel University London. Her research focuses on developing functional engineered tissues for applications in tissue regeneration, drug testing, and disease modelling, with an additional interest in organ-on-a-chip engineering. Her research group aims to create functional engineered tissue platforms by developing novel biomaterials and biofabrication techniques, integrating 3D printing and drug-loading approaches to enhance the functionality of engineered tissues and organoids for repairing damaged tissues and organs, such as wound healing and bone regeneration.

Amirpasha Moetazedian

Amirpasha Moetazedian is an assistant professor in Medical Engineering at the School of Engineering and Technology at the University of Hull. He is the programme director of the Biomedical Engineering course at the University of Hull. His research activity is focused on biomanufacturing and biomaterials for tissue engineering and regenerative medicine. Specifically, his research group is involved in the development of novel microfluidic devices for 3D bioprinting, wearable sensors and in vitro modelling. He is currently the co-lead of a £2.9M MRC-funded project ARTEMIS for the development and validation of the next generation of artificial blood vessels for cardiovascular diseases.

Peter Zioupos

Peter Zioupos is Emeritus Professor in Biomedical Engineering at the School of Engineering and Technology at the University of Hull. He has held previous appointments at the universities of Cranfield, York, and Strathclyde. He has an interest in Musculoskeletal science and research and their link to Orthopaedic Biomechanics, Bone physiology, Anatomy and Forensics. Professor Zioupos is also active in forensics and offers consultancy in legal cases regarding lethal or otherwise trauma, which involves primarily bone injuries. He has served as a Vice-President at the Council of the European Society of Biomechanics, and is currently the Vice-President and Trustee of the Chartered Society of Forensic Sciences.

Preface

"Recent Advances in 3D Printing Technologies in Bioengineering" is a reprint of a Special Issue in the *Journal of Manufacturing and Materials Processing*, which contains research articles in the rapidly evolving field of 3D printing/additive manufacturing (AM) for the fabrication of functional materials with complex geometries which offer transformative applications in bioengineering. These papers deal with cutting-edge technologies which integrate material design, structural control, biological functionality and future clinical strategies, which position AM at the forefront of modern biomedical innovation.

Bin Zhang, Amirpasha Moetazedian, and Peter Zioupos
Guest Editors

Journal of
Manufacturing and Materials Processing

MDPI

Article

Spine Motion Segment Analogues: 3D Printing, Multiscale Modelling and Testing to Produce More Biofidelic Examples

Constantinos Franceskides [1], Tobias Shanker [1], Michael C. Gibson [1] and Peter Zioupos [2,*]

[1] Cranfield Forensic Institute, Cranfield University, Cranfield MK43 0AL, UK
[2] Biomedical Engineering, School of Engineering, University of Hull, Kingston upon Hull HU6 7RX, UK
* Correspondence: p.zioupos@hull.ac.uk

Abstract

Computed tomography and magnetic resonance imaging are two powerful modalities which can be used in the clinical setting to produce data for the creation of patient-specific finite element analysis (FEA) models and physical analogues—for instance, by using additive manufacturing (AM)—that mimic the properties of soft and hard tissues, both morphologically and mechanically. However, there remains a gap between creating a perfect biofidelic physical analogue and its computational counterpart. This gap exists because, firstly, in silico models are often too complex to realise, and secondly, real-life conditions are challenging to emulate both computationally and mechanically, as they involve multiscale situations that are inherently heterogeneous and patient specific. In this study, we applied a multi-scale approach to design and model porcine vertebral specimens. Our results identified critical design factors that affect the quality and accuracy of the models, specifically highlighting that scanning resolution/fidelity and the thresholding technique have a directly proportional impact on model accuracy. A small shift up and down the greyscale level by 20 units can affect the behaviour of the AM sample by as much as [−15% +47%]. Working up the levels for manufacturing, testing and modelling (i) cylindrical cores to (ii) whole vertebrae and then (iii) a whole spine motion segment, we observed that the fidelity of predictions reduces, and errors increase as the structure becomes more complicated and intricate (3.6%, 7.5% and 15%, respectively). We are confident that further material-level developments will provide solutions for the more intricate parts of spinal motion segments, such as the intervertebral discs and facets, which in their natural form are highly sophisticated structures. To the best of our knowledge, this is the first time a holistic multiscale approach has been implemented to produce AM biofidelic analogues of skeletal parts. Our data showed good agreement between the physical and in silico models, confirming that it is possible to model real-time objects and situations both physically and in silico. This ultimately will enable the development of accurate, patient-specific physical models for use in biomechanical testing and medicolegal applications.

Keywords: clinical scanning; image-based biomechanics; patient-specific FEA; multiscale modelling; additive manufacturing; biofidelic physical analogues

1. Introduction

The combination of additive manufacturing and increasing computational power has introduced new concepts in the development of biomechanical analogues and corresponding models, allowing for the creation of robust in silico and physical models that overcome the complex ethical procedures concerning the use of cadaveric specimens. Rapid prototyping and 3D printing has proved invaluable in biomedical engineering, orthopaedic surgery and clinical education [1–6], and in forensic applications by offering physical size models which can be handled and used to demonstrate structure and materials [7,8]. Finite element analysis, in particular, has facilitated the characterisation and testing of different components of the musculoskeletal system [9]. This evolution in computational and screening techniques has driven research towards complex and unknown territory [10,11], where bone macro-modelling is increasingly replaced by highly detailed and resource heavy micro-scale models [12]. The creation of biofidelic models, whether in silico or in real life, is more complex when investigating intricate geometries where the interactions between different subcomponents (such as soft and hard tissues) play a key role [13]. The reliability of such models depends heavily on how they are defined, interpreted and implemented for each application. These definitions are based on approximations and compromises limited by technical and resource constraints such as 3D printing resolution, scanning capability and the trade-off between computational speed and accuracy [10,12–14].

1.1. Recent Applications—Skull

We have recently used a combination of 3D printing and FEA (finite element analysis) modelling to study skull fractures in impact [15,16]. The results were remarkably reproducible and offered insight in fracture scenarios, where if one used real tissue samples, every test would yield a different outcome. By producing many copies of the same skull (reproducing the geometry accurately with exactly the same material each time), it is possible to observe if a fracture pattern, fracture being a stochastic event, is a result of the bone material heterogeneity, the bone structure, or both. Two aspects of the skull fracture were particularly revealing: firstly, the cracks emanating from the impact site were always of the same pattern and directionality in every test, indicating that the crack progress had more to do with the structure and geometry of the skull than what it was made of, and secondly, smaller cracks at the base of the skull were at the same site in the various tests and also where FEA models were predicting them to be (Figure 1).

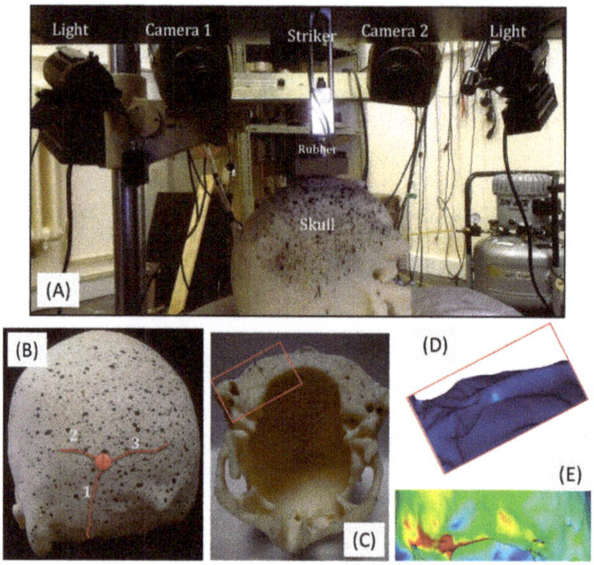

Figure 1. An example of a holistic approach involving the creation of both a physical model and an in silico (computational) model of a human skull, as studied in [15,16]. (**A**) A number of replicas of the

https://doi.org/10.3390/jmmp10020056

skull were impacted at the same point in a drop tower, suitable illumination and high speed cameras were used, full details in [15,16]; (**B**) crack patterns at the site of impact were emanating in 3 diverging crack routes, which were identical in all replicas of the skull used [15]; (**C**) other cracks at the base of the skull (red square) appeared reliably and repeatedly in certain areas and where FEA modelling predicted them to appear (enlarged views of the area in (**D**,**E**)), showing a synergy and agreement between these two powerful biomechanical testing methodologies (AM produced physical model and in silico computational model).

1.2. Recent Applications—Vertebrae

In spine biomechanics, motion segment analogues have been produced with varying levels of success [17,18]. Outsourcing the development of custom analogues requires several months of lead time and involves prohibitive costs and ethical restrictions, which limit their accessibility. These analogues are produced mostly for demonstration and training purposes, and as such do not fully simulate the mechanical behaviour of the natural parts they represent. Biofidelic copies of vertebrae and spine motion segments have been validated by quasi-static compression loading in vitro [19]. 3D-printed analogues were created using readily available and off-the-shelf materials with known mechanical properties: namely, acrylonitrile butadiene styrene (ABS) for the vertebral body and a two-part liquid polyurethane rubber system (of an appropriate Shore hardness value) for the intervertebral disc (IVD) and the superior and inferior process ligaments [19–21].

1.3. The Need for Multilevel Validation

The present study explores the potential of finite element analysis (FEA) for the modelling and validation of 3D-printed spinal analogues under quasi-static loading (0.01 mm/s), which is investigated using a bottom-up approach (from small sample size and increasing the length scale) instead of the top-down strategy applied in other multiscale studies [22,23]. This would allow us to determine the lowest fidelity needed to provide an effective method for the validation of the physical model, thus reducing the costs and lead time for a patient-specific FEA solution. A biofidelic finite element model must be fed with material properties in order to calculate accurate stresses and strains. The assignment of these properties at the microscopic scale is derived from measurement at this level, which usually involves a surrogate physical variable such as the micro-CT greyscale (GS). This is dependent on the linear attenuation coefficient of the material, which is linked to the material density.

In order to generate a material model for ABS over a wide applicable range (micro to macro), there needs to be a discernible variation in the average GS values between samples. Here, we achieved this in three steps by testing specimens with (1) longitudinal holes in solid cores (longitudinal cores), (2) solid, high-density and low-density filled cores (transverse cores), and (3) artificial vertebral cores (AVCs) with varying threshold levels. The most accurate material model was selected and applied to higher-scale segments: (1) an artificial vertebral body (AVB) and (2) an artificial motion segment (AMS), comprising two vertebrae with the intervertebral disc (IVB) in between, as described in our previous work [20,21]. We have thus investigated the best practices for generating an FEA model of a physical analogue that is suitable for multiscale analysis, ultimately aiming to facilitate the development of patient-specific tissue (and bone tissue, for that matter) analogues which can be used in orthopaedic and biomechanical applications.

2. Materials and Methods

All samples (whole vertebral bodies and cores) were printed using a fused deposition modelling (FDM) UPrintSE printer (Stratasys Inc., Eden Prairie, MN, USA), based on computer aided design (CAD) geometry samples at 100 μm, a scale that achieves mechanically

https://doi.org/10.3390/jmmp10020056

accurate 3D-printed replicas of biological objects [15,20]. Tool paths for 3D printing were generated using CatalystEX v4.5 (Stratasys Inc., Eden Prairie, MN, USA). Production of whole AVB samples were based on micro-CT image stacks obtained from six porcine vertebral bodies. The entire vertebral body was sectioned, removing the endplates, vertebral arch and pedicles. The AMS analogue was based on RMS#4 (Real Motion Segment), one of these six spine specimens, and was created using the same method as the AVB samples, with one exception: IVD and facet joints were introduced that fused the top and bottom vertebral bodies. The IVD and intra-facet cartilage (area between facet joints) were formed from PT Flex 85 liquid rubber of Shore A hardness 85 (Polytek Development Corp., Easton, PA, USA). The two-part rubber was mixed and then injected into a cast built around the IVD superior and inferior endplates and the superior and inferior articular processes, using Sugru mouldable glue (FormFormForm Ltd., London, UK). Accurate IVD thickness (based on CT reconstructions) was ensured by placing struts that were equal in height to the IVD (also produced using PT Flex 85) at three points on the endplates. Figure 2 shows the completed sample following the injection of the IVD and facet ligaments, which has been reported in detail by Franceskides et al. [20,21].

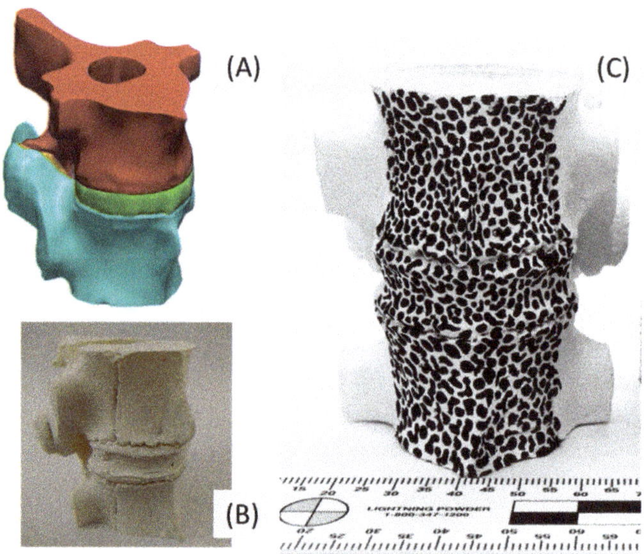

Figure 2. Artificial motion segment, as used and analysed in a previous study [19]; (**A**) surface model following CT scanning ready for use with AN; (**B**) 3D-printed sample as a physical model; (**C**) AM physical model with a surface speckle ink pattern to allow surface strain mapping by using DIC.

Core samples with longitudinal holes (Figure 3A) were created using SolidWorks 2018 SP1 (Dassault Systèmes SolidWorks Corporation, Vélizy-Villacoublay, France). For these samples, the diameter of the hole was kept constant at 2 mm, but the number of holes varied (0, 1, 3, 5 and 7).

The transverse cores with varying fill distribution (Figure 3B) were created by modelling a solid cylinder and using CatalysEX to select three different default fills: solid, high density or low density. Using these settings, the external structure was printed as a solid, whereas the density of the internal structure was varied by controlling the amount of fill, achieving a more even density variation throughout the cylinder (particularly in the x and y planes) compared to the longitudinal cores. Both the longitudinal and transverse cores were 9 mm in diameter and 25 mm long to maintain comparability between the test results.

 https://doi.org/10.3390/jmmp10020056

The data from the longitudinal and transverse cores were pooled (LTCs) for broader density variation and to improve the population size in comparison to the AVCs.

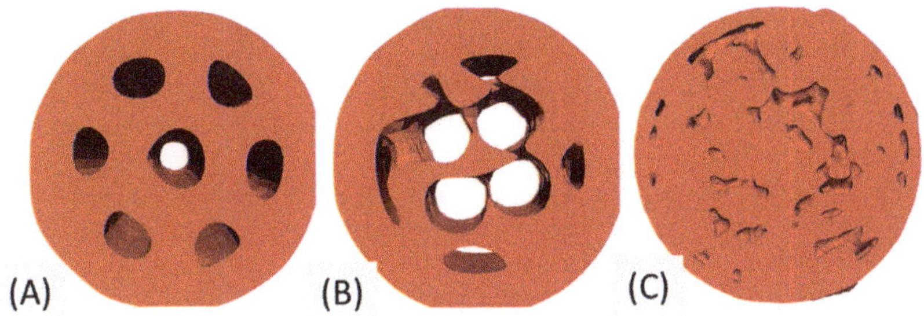

Figure 3. Examples of (**A**) longitudinal, (**B**) transverse and (**C**) artificial vertebral cores.

The AVCs were generated by segmenting micro-CT scans from two cadaveric porcine motion segments (real motion segments, named RMS#4 and RMS#5) obtained in our previous studies [19–21]. These were derived from two porcine vertebrae at levels L1-L2 and T12-L1, respectively (Figure 3C). The cores were taken from a cylindrical region of interest, 9 mm in diameter, and aligned with the axis of loading and axis of the spine. The lengths were limited by the size of the actual vertebrae (without the endplates), so that AVCs were shorter than the longitudinal and transverse cores (LTC). Core density was varied by controlling the amount of bone thresholded compared to the baseline. The thresholding level has a significant effect on the microarchitectural values produced for cellular solids, as has been shown recently [24]. Models with a threshold grey scale value of ± 20 in relation to the baseline were generated in addition to the standard upper/lower peak thresholding procedure. Figure 4 shows an example of the segmentation. A thin solid shell of material was added to the circumferential edge of the cores (to create a solid surface around the core), as shown in blue, to prevent tool path generation errors, which are generated by the cellular discontinuous geometry of these cellular samples.

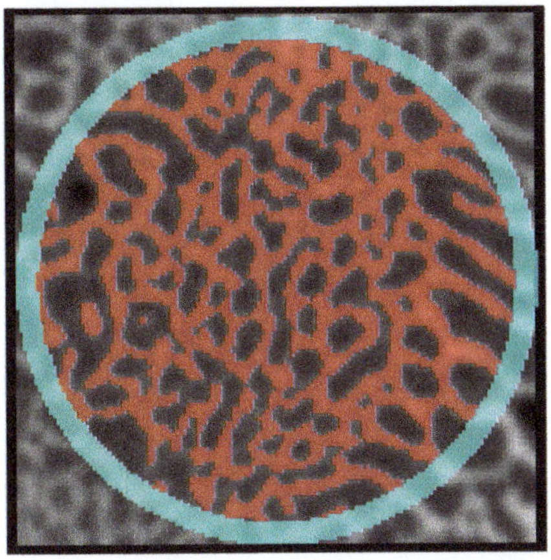

Figure 4. Artificial vertebral core segmentation with an outer shell added circumferentially around the edge (The bright colours are only used for illustration of the segmentation).

The AVCs were bathed, prior to scanning, in a Support Cleaning Apparatus (PADT Inc., Tempe, AZ, USA). This ensured that all the internal support was removed without damaging or affecting the rest of the structure. Once all the support was removed, samples

 https://doi.org/10.3390/jmmp10020056

were scanned at the same resolution and settings as stated above for the longitudinal and transverse cores.

These were scanned using a CT H225 cone beam μCT scanner (Nikon Metrology UK Ltd., Tring, UK) operated at 80 kV and 65 μA. Noise reduction and beam hardening correction were achieved using CT Pro 3D (Nikon Metrology UK Ltd.). The geometric magnification produced a voxel dimension of 26 μm for all the cores, 55 μm for the AVBs and 89 μm for the AMS experiment. The data generated during the scans were used for finite element modelling and morphometric analysis. ImageJ v1.52 [25] with the BoneJ v1.4.3 [26] plugin was used to calculate the GS values of the ABS peak and air peak, as well as the volume fraction, degree of anisotropy and mean GS through the entire cylinder. Table 1 shows the volume fraction and weight of the LTC samples.

Table 1. Volume fraction and weight of the longitudinal and transverse cores (LTCs).

	Samples	Volume Fraction	Weight (g)
1	Solid	0.999	1.507
2	0 Holes	0.998	1.503
3	1 Hole	0.916	1.386
4	High Density	0.907	1.301
5	3 Holes	0.843	1.290
6	5 Holes	0.773	1.168
7	7 Holes	0.736	1.062
8	Low Density	0.624	1.006

Table 2 shows the allocation of cores from the original specimens, highlighting parameters such as the volume fraction, weight, volume and length.

Table 2. Properties of artificial vertebral cores.

Motion Segment	Vertebral Body	Core Section	Volume Fraction	Mass (g)	Volume (mm^3)	Length (mm)
RMS#4	RVB4_T	4BN	0.787	1.049	1059.22	20.60
		4BO	0.568	0.831	802.08	20.26
		4BU	0.929	1.232	1217.84	20.52
	RVB4_B	4TN	0.836	1.101	1066.75	19.87
		4TO	0.760	0.994	978.78	19.89
		4TU	0.920	1.208	1183.93	19.85
RMS#5	RVB5_T	5BN	0.715	0.880	842.64	17.98
		5BO	0.594	0.760	727.21	17.98
		5BU	0.817	0.996	972.45	17.98
	RVB5_B	5TN	0.722	0.890	872.23	17.34
		5TO	0.696	0.828	795.43	17.05
		5TU	0.830	0.961	916.46	17.36

2.1. Mechanical Testing

All analogues were tested under compression in a Dartec Series HC25 servo-hydraulic testing machine (ZwickRoell Ltd., Worcester, UK) with a 9610 controller running its own Workshop 96 software. The tests were conducted from rest to plastic failure (or force limit of the load cell, whichever came first) under crosshead displacement control at 0.01 mm/s. The forces were logged using a 5-kN load cell (Sensotec, RDP Electronics Ltd., Wolverhampton, UK). The core specimens were held centred in endcaps. The bottom endcap was fixed, whereas the top endcap featured a self-levelling platen (Figure 5F).

 https://doi.org/10.3390/jmmp10020056

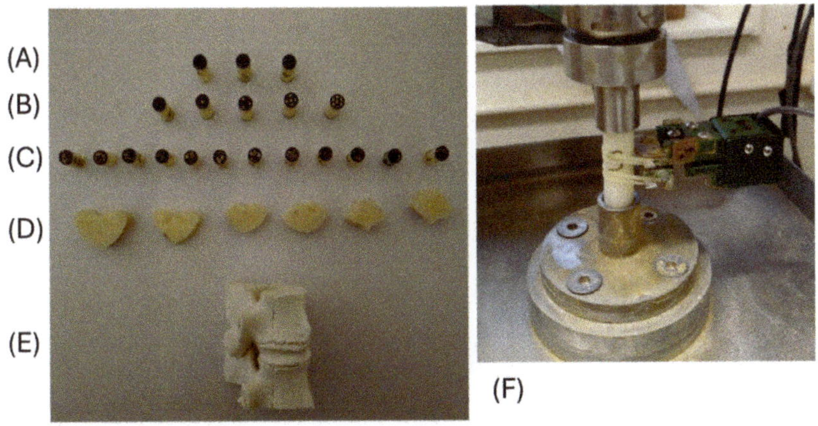

Figure 5. Porcine spinal analogues. Starting from the topmost samples: (**A**) three transverse cores of different densities, produced by setting the printer to use infills of solid, low density and high density; (**B**) five longitudinal cores with varying number of holes (practically same in-fill setting but in effect variable apparent density, depending on the number of holes present: 0, 1, 3, 5, 7 from left to right); (**C**) twelve transverse cores produced by varying the threshold levels of the scans; (**D**) six vertebral bodies; and (**E**) a complete AM motion segment. (**F**) Mechanical testing setup of the cores with a contact extensometer attached and held in place against the central portion by rubber bands.

In the AMS experiment, the loading was delivered using a flat, spherical-joint, self-levelling platen at the top and a fixed, flat platen at the base, whereas the AVBs were fixed between an endcap, spherical-joint platen at the top and a fixed-endcap platen at the base.

2.2. Finite Element Model

Geometry: The CT images were segmented using Simpleware ScanIP M-2017.06-SP2 (Synopsys Inc., Mountain View, CA, USA). CT scans were resampled to 100 µm for consistency between meshes. The outer surface of all samples was identified using standard thresholding techniques. The meshes were created using quadratic 10-node tetrahedral elements in the FE-Free meshing routine of Simpleware ScanIP. This was done at the coarsest setting (−50) to reduce the computational cost of the models as much as possible, allowing for analysis on the less capable machines that are often found in a clinical environment. The average GS air peak and the measured apparent density (mass/volume) [27] were used to draw a linear relationship between the GS and apparent density (Figure 6).

Table 3. ANSYS mechanical properties of idealised and artificial vertebral cores (AVCs). ABS: Acrylonitrile butadiene styrene and IVD: intervertebral disc.

Material Type	GS-Density Relationship (mg/mm^3)	Elastic Modulus (MPa)	Poisson's Ratio
ABS (idealised geometry)	3.098×10^{-11} GS − 7.15×10^{-12}	$1.244 \times 10^{19}\ \rho^{1.752}$	0.35
ABS (AVCs)	3.082×10^{-11} GS + 1.31×10^{-11}	$4.135 \times 10^{25}\ \rho^{2.489}$	0.35
IVD	N/A	2.5	0.49

https://doi.org/10.3390/jmmp10020056

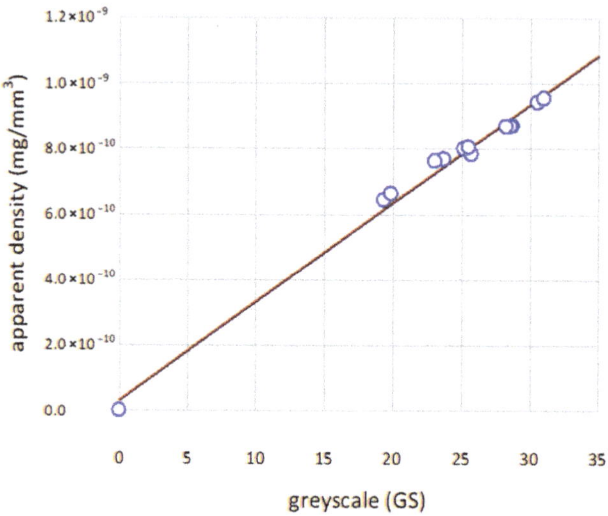

Figure 6. Artificial core apparent density plotted against grey scale (GS). Model curve (red line) App Dens = 3.082×10^{-11} GS + 1.31×10^{-11} (Table 3).

The relationship between density and elastic modulus was described using a fitted power law curve (Figure 7) that was similar to that reported for trabecular bone samples [28].

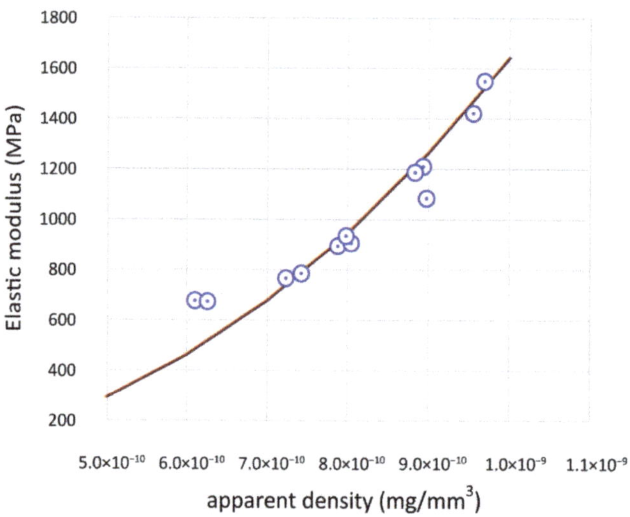

Figure 7. Artificial core elastic modulus plotted against apparent density. Model curve (red line) $E = 4.135 \times 10^{25}\ \rho^{2.489}$ (Table 3).

This was done twice: once for the LTC and again for the AVC samples. A Poisson's ratio of 0.35 [29,30] was applied to complete the linear elastic model for the bone analogues. The AMS included an additional analogue two-part rubber IVD, which was modelled as a human intervertebral disc with an elastic modulus of 2.5 MPa and a Poisson's ratio of 0.49. This value is representative of the entire composite body, comprising the nucleus pulposus and annulus fibrosus: the two ground substance inner and outer components of the IVD [31]. Two FE material models were created, one (FLC) based on the 3DP samples

https://doi.org/10.3390/jmmp10020056

with the geometry of idealised longitudinal cores (computer designed cancellous inner architecture) and another (F-DLS) on the AVCs (scanned real cancellous architecture). Their relationships and material properties are compared in Table 3.

The models generated for all cores (LTCs and AVCs) comprised ~15,000 elements and were analysed using ANSYS APDL v15.0 (ANSYS Inc., Canonsburg, PA, USA) with an implicit solver. The top face was coupled to replicate the effects of a platen and displaced by 1 mm, whereas the bottom face of each specimen was fixed in all directions. The nodal reaction forces on the top face were computed to provide a stiffness measurement for validation against the mechanical testing.

3. Results

Accurate patient-specific finite element models and their analogues are needed in the fields of forensic and injury biomechanics, as well as orthopaedics. Micro-CT data were used to develop analogue models of porcine spinal motion segments with subsequent FEA models. These were then tested in a similar manner to compare their stiffness and reaction forces. Clinical CT scanning is currently conducted at 300–500 μm [32], which is a much lower resolution than the image segmentation used for the FEA in this study (100 μm). The present scans produced a more accurate FEA model because there was more data upon which to base the model. The discrepancy between the in vivo and in vitro CT scanning resolution could be a limiting factor for the clinical application of this model.

3.1. Artificial Cores

The arithmetic mean and standard deviations for the morphological results of the samples are shown in Table 4.

Table 4. Comparison of morphometric data for artificial vertebral cores (AVCs) and longitudinal and transverse cores (LTCs), with asterisks (*) indicating significant differences.

	LTC		AVC		AVC Sub	
	Mean	**Std. Dev**	**Mean**	**Std. Dev**	**Mean**	**Std. Dev**
Length (mm)	25.15 *	0.132	18.89 *	1.321	18.95 *	1.331
Volume fraction	0.849	0.1233	0.765	0.1082	0.765	0.0496
D_{app} (Mg/mm^3)	7.96×10^{-10}	1.07×10^{-10}	8.12×10^{-10}	9.27×10^{-11}	8.12×10^{-10}	3.71×10^{-11}
DA	0.987 *	0.0157	0.305 *	0.0925	0.305 *	0.0248

To compare the morphometry with that of porcine trabecular bone, the four AVCs were thresholded using the standard method (4BN, 4TN, 5BN, 5TN) and then subsampled into AVC Sub; two-sample *t*-tests were used to calculate significant differences in relation to LTCs ($p = 0.05$). The AVC and AVC Sub samples featured virtually indistinguishable means; therefore, in this context, they can be used interchangeably. The length was consistent for the LTCs, but there was greater variation in the AVC and AVC Sub samples. The volume fraction was 8% higher on average for the LTCs, but showed greater variance than AVC and AVC Sub. The apparent density of the two models was similar and showed an inter-specimen correlation when normalised against the volume fraction (6.76×10^{-10} LTC, 6.21×10^{-10} AVC). The LTC dataset was mostly orthotropic, whereas the AVC dataset showed a much more anisotropic construction (0.305). Figure 8 shows the variance in the experimental stiffness values for the different vertebral bodies and its dependence on the thresholding method, revealing a nonlinear relationship between the −20 and +20 GS.

https://doi.org/10.3390/jmmp10020056

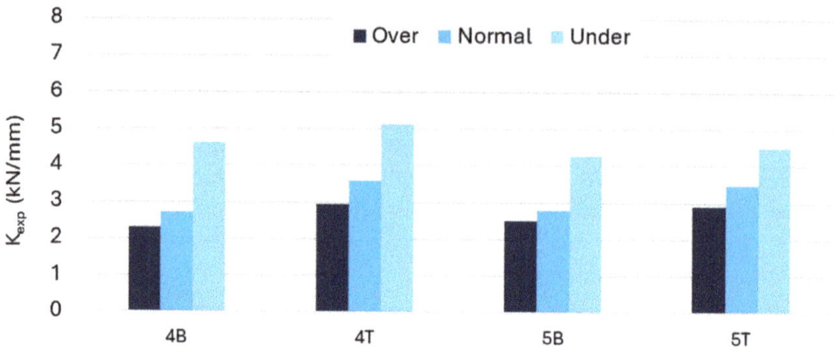

Figure 8. Experimentally obtained stiffness (K_{exp}) values for the 3 different thresholds on 4 samples. When the threshold value increased (less material in the 3DP sample), the material stiffness decreased by 15.3% on average (average DK = −0.48 kN/mm; significant paired *t*-test $p = 0.009 < 0.05$). For a similar shift in the threshold to the lower value (added more material in the 3DP sample) and the stiffness increased on average by 47.3% (average DK = +1.49 kN/mm; significant paired *t*-test $p = 0.004 < 0.05$), showing that the thresholding level had a significant and non-linear effect on the properties of the produced 3DP sample.

The morphometric data shown in Table 4 reveal several trends. The AVC and AVC Sub data have similar arithmetic means, but the AVC Sub data have a much smaller standard deviation. This is expected, because the population has been reduced from 12 to four but is segmented in the same manner, reducing the morphological variance. It also shows that different thresholding methods have a linear effect on morphological measurements. This is in contrast to the nonlinear mechanical relationship between the thresholding techniques (Figure 8).

The apparent density is controlled by the threshold value applied during segmentation. Our model shows a nonlinear relationship between the apparent density and experimental elastic modulus, which is reflected in the power exponent in the fitted material model (${\rho_{app}}^{2.489}$, $R^2 = 0.944$), which is in the range of previously reported bone FEA models (${\rho_{app}}^{1.49}$–${\rho_{app}}^{3}$) [12].

The degree of anisotropy for the LTCs was close to 1 ± 0.016 (mean ± SD), which is understandable, as these are computer-designed and have directionality. Meanwhile, the AVCs had anisotropy of 0.305 ± 0.0925, which was similar to previously reported porcine vertebral anisotropy values of 0.321 ± 0.067 [33]. There was no significant difference in apparent density between the LTC and AVC samples (two-sample *t*-test, $p = 0.754$) and when normalised against the volume fraction, the mean values were much closer (6.76×10^{-10} LTC, 6.21×10^{-10} AVC), showing that there is little variance in the 3D-printing procedure, which could affect the mechanical results. Furthermore, no significant differences in volume fraction were observed between the cores. Therefore, both sets had a similar spread of volume fractions, indicating that other morphological properties affect the mechanical response and model characteristics.

Bone has a higher elastic modulus than ABS, 10–20 GPa vs. 3 GPa, respectively, so a greater volume of ABS material (higher BV/TV inevitably) is needed to achieve a mechanically comparable behaviour at the macrostructural level [19]. As previously recommended [34], we measured the bone volume divided by the total volume (BV/TV) of porcine vertebrae in the range 0.30–0.40, whereas the cores used in this study were 0.80 ± 0.12 overall and the AVBs were 0.82 ± 0.98. Given the lower elastic modulus of ABS, higher BV/TV values are necessary in bone analogues and are therefore adjusted so that the models achieve a similar stiffness upon loading [19].

https://doi.org/10.3390/jmmp10020056

The in silico and experimental stiffness results for the longitudinal and transverse cores showed good agreement. We obtained two computational stiffness values: one using model F-DLS and the other using model FLC (Table 5).

Table 5. Longitudinal and transverse core stiffness results. K: Stiffness. Exp: Experimental value. FEA: Finite element analysis value. F-DLS and FLC are FE models.

Sample	Volume Fraction	Weight (g)	K_{EXP} (kN/mm)	$K_{(F\text{-}DLS)}$ (kN/mm)	$K_{(FLC)}$ (kN/mm)
Solid	0.999	1.507	3.917	3.710	4.978
0 Holes	0.998	1.503	3.084	2.909	4.997
1 Hole	0.916	1.386	3.569	3.339	4.036
High Density	0.907	1.301	3.113	3.020	3.947
3 Holes	0.843	1.290	2.623	2.262	4.587
5 Holes	0.773	1.168	4.044	3.945	4.167
7 Holes	0.736	1.062	2.975	2.818	3.311
Low Density	0.624	1.006	2.294	1.692	2.482

Model F-DLS under-predicted stiffness by 10.5% on average (average DK = −0.24 kN/mm; significant paired t-test $p = 0.005 < 0.05$), whereas model FLC over-predicted the stiffness by 26.1% on average (average DK = +1.10 kN/mm; Significant paired t-test $p = 0.004 < 0.05$). Figure 9 compares the computed stiffness values of the two FE material models against the experimental values for the longitudinal cores; the more significant qualitative result is that the behaviour of the FLC models was erratic, while the F-DLS models followed the experimental data linearly and proportionately.

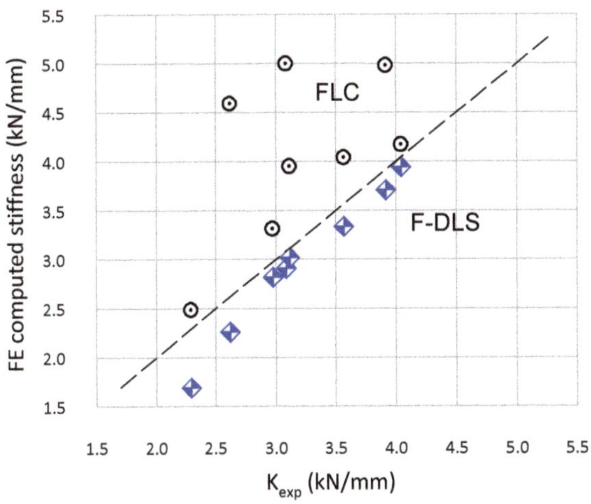

Figure 9. FE model stiffness values for FLC (circles) and F-DLS (diamonds) models vs. experimentally determined stiffness (K_{exp}). The line of equality is also shown (dashed).

Table 6 summarises the stiffness results for AVCs and shows a 3.6367–371 average under-prediction of stiffness (average DK = −0.13 kN/mm; not significant paired t-test $p = 0.121 > 0.05$) when comparing the F-DLS material model to the experimental data. By contrast, the FLC models over predicted the stiffness values by 55.8% on average (average DK = +1.83 kN/mm; significant paired t-test $p = 0.001 < 0.05$). Like previously, while the F-DLS models followed the experimental data linearly, the FLC models were erratic (presenting noise). Figure 10 compares the computed stiffness of the two material models against the experimental values for the cores created using RMS#4 and RMS#5.

 https://doi.org/10.3390/jmmp10020056

Table 6. Artificial vertebral core stiffness results. BV/TV: Bone volume divided by the total volume. K: Stiffness. Exp: Experimental value. FEA: Finite element analysis value. F-DLS and FLC are material models.

Sample	BV/TV	K_{EXP} (kN/mm)	$K_{(F\text{-}DLS)}$ (kN/mm)	$K_{(FLC)}$ (kN/mm)
4BN	0.787	2.725	2.879	4.525
4BO	0.568	2.313	2.172	2.821
4BU	0.929	4.635	5.467	6.973
4TN	0.836	3.59	3.636	5.253
4TO	0.760	2.947	3.029	4.710
4TU	0.920	5.127	4.948	6.314
5BN	0.715	2.79	2.973	4.773
5BO	0.594	2.496	2.538	5.364
5BU	0.817	4.268	4.276	6.107
5TN	0.722	3.474	3.433	5.435
5TO	0.696	2.898	3.276	5.128
5TU	0.830	4.500	4.701	6.280

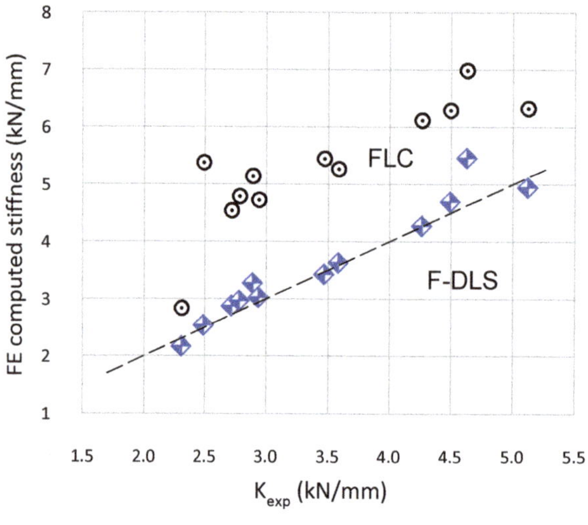

Figure 10. AVC4 and AVC5 core F-DLS (diamonds) and FLC (circles) material models. FEA-derived stiffness (K_{FEA}) plotted against experimentally determined stiffness (K_{Exp}). The line of equality is also shown (dashed).

3.2. Artificial Vertebral Bodies

Table 7 and Figure 11 shows morphometric and mechanical testing data from the AVBs, using the F-DLS material model. This repeats the previous analysis, now at a higher level (vertebrae are larger and more intricate in geometry than simple cores) by focusing on the use of the more successful F-DLS model. The FE models consistently over-predicted the stiffness of whole vertebral body samples by 7.5% on average (average DK = +1.83 kN/mm; not significant paired *t*-test $p = 0.014 < 0.05$), but followed the experimental data with fidelity and in a linear manner.

 https://doi.org/10.3390/jmmp10020056

Table 7. Artificial vertebral body morphometric data and experimental results. BV/TV: Bone volume divided by the total volume. K: Stiffness. Exp: Experimental value. F-DLS is a material model.

Body	Mass (g)	BV/TV	V (mm^3)	Rho (Homo)	K_{exp}	$K_{(F\text{-}DLS)}$
AVB1	5.942	0.825	6.67×10^3	8.91×10^{-10}	24.345	27.704
AVB2	3.283	0.739	3.94×10^3	8.34×10^{-10}	18.946	20.106
AVB3	3.046	0.845	3.13×10^3	9.72×10^{-10}	27.615	30.646
AVB4	3.966	0.822	4.30×10^3	9.21×10^{-10}	19.268	21.339
AVB5	3.250	0.826	3.62×10^3	8.97×10^{-10}	17.003	17.365
AVB6	4.137	0.845	4.61×10^3	8.98×10^{-10}	17.561	18.528

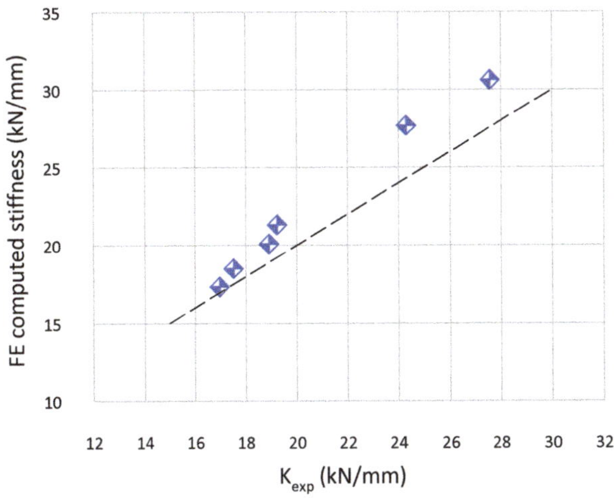

Figure 11. Behaviour of AVB data (artificial vertebral bodies): FEA-derived stiffness $K_{(F\text{-}DLS)}$ vs. experimentally determined stiffness K_{exp}. The line of equality is also shown (dashed).

3.3. Artificial Motion Segments

The F-DLS material model agreed well with the experimental dataset, providing confidence at this stage to progress the analysis to the next higher level, which is the artificial motion segment (AMS). AMS comprises two vertebrae and the intervertebral disc (IVD) between them. No further tests were conducted with the FLC model, due to its relative inaccuracy compared to the F-DLS model. At this point, we concentrate on the actual force/displacement curve produced in tests and in silico.

The reaction forces recorded for the AMS analogue are compared to the corresponding FEA model tested using the F-DLC material model in Figure 12. An AMS comprises two 3DP vertebrae of ABS and a rubberlike artificial IVD. The non-linear properties of the IVD part of the manufactured physical model only allow for the coincidence of test results and model predictions over a range of displacements, which can only be tuned by the researcher within a range, as shown here for displacements above 1.3 mm. We cannot elaborate any further on the suitability of the two-part liquid polyurethane rubber system (PT Flex 85, Polytek, Easton, PA, USA) we used here, for producing a more biofidelic intervertebral disc. The solution we used was tuned for the right Shore hardness for the physical model and a modulus of 2.5 MPa for the in silico one. Of course, there are other physical properties and the very linearity of the curve that must be accurately matched to the IVD material. This is beyond the scope of this paper. It is very likely that with the new technological advancements in 3D printers, which can inject a variety of softer polymers, an artificial IVD can be produced that possesses more biofidelic properties and a composite hierarchical organisation and microstructure.

https://doi.org/10.3390/jmmp10020056

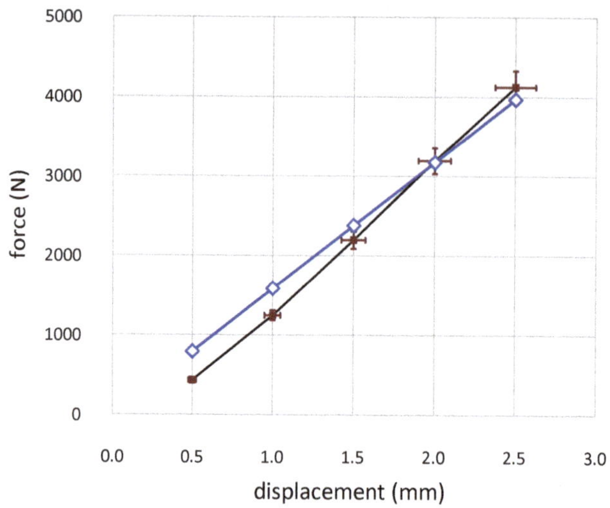

Figure 12. Behaviour of the AMS data: comparison of experimental (red) and computational force/displacement data (blue). (Experimental error bars were on average ±5% for load and 8% for displacement) The computational and experimental results agreed closely (NS difference) for displacements of 1.3 mm and above. The experimental data contain the non-linear nature of the artificial IVD material and other real-world phenomena, such as contact effects with the platens where there are asperities, and these show as an increase in stiffness as the load increases. The computational data of course has no such end contact effects, as all loading scenarios are idealised in silico.

The IVD properties played a greater role in determining the overall stiffness response of the motion segment than expected, both for cadaveric and analogue samples, as suggested in earlier reports [35]. This is because vertebral bodies are orders of magnitude stiffer than the IVDs and associated soft-tissue ligaments, leading the IVDs to deform more readily and fail earlier under compressive loading than the vertebral bodies. This highlights a greater need for accurate IVD and soft tissue replication in analogues and the models based on them. When generating the AMS finite element model, the IVD and facet ligaments were created using the faded disc boundary, which may have overestimated or underestimated the motion segment stiffness, depending on how the soft tissue was reconstructed, leading to the differences in observed stiffness.

Overall, the results of this study indicate that with a careful, staged and multiscale approach where one gradually increases the degree of complexity, good agreement between the physical and in silico model can be achieved. This promises to deliver both physical and in silico models which mimic real musculoskeletal structures well, and this offers a range of opportunities for scientists, practitioners and manufacturers in terms of implant design and applications.

4. Discussion

The present paper works around the research hypothesis that a multiscale modelling approach, which integrates high-resolution imaging data (such as CT/MRI), optimised thresholding techniques, and additive manufacturing, can produce biofidelic physical and computational models of skeletal structures (e.g., porcine vertebrae) that accurately replicate both morphological and mechanical properties, enabling reliable use in clinical and biomechanical applications [1–6]. The outcomes of this approach can be used in pre-surgical planning and simulation; biomechanical testing for research; medical education, allowing

 https://doi.org/10.3390/jmmp10020056

for hands-on approach methods [3–5]; medicolegal reconstruction (forensic biomechanics cases [7,8]) and the design and development of implants and surgical tools [2,3,5].

The present paper describes a clinical problem that is centred around the challenges of accurately replicating human (or animal) anatomical structures and biomechanical behaviour, using both computational (FEA) and physical (AM) models for clinical and research purposes. The challenges arise from the inherent limitations of both these two methodologies (FEA and AM) in imaging resolution, modelling complexity and material fidelity. More specifically, our study reinforced and added to the well-known challenges which mainly have to do with (i) computational models (in silico) that are too complex and/or idealised; (ii) real world biological variability and mechanical behaviours that are hard to replicate; (iii) imaging quality (CT/MRI resolution and thresholding) which can significantly influence model accuracy; and (iv) soft tissue modelling, especially for the complex spinal components we examined here, which are limited by material and manufacturing constraints [2,6,10,13,23,35].

In applying our two-modalities (FEA/AM) methodology on a spine section as an example, we reaffirmed that biological tissues are hierarchically complex and the attempts of modelling them (physically and computationally) are challenging at multiple length scales [15,16,19–21]. For spine segments, in particular, we saw that the intervertebral discs (IVDs) are difficult to model, which is consistent with known gaps in spinal biomechanics research. We could also clearly see that the well-known trade-off between radiation dosage and CT imaging resolution is a problem which is here now and must be addressed in the future. However, the arrival of new high-resolution in vivo scanners (achieving 300 micron in vivo) allow for the same scan data to be used for both FEA modelling and 3DP. A recent paper from our lab demonstrated the production of FEA models from such modern clinical scanners and the computational fidelity that can be expected in such applications [36]. In carrying out the present study, it was clear to us that there are existing gaps and challenges in the literature and acquired biomechanical knowledge due to the lack of standard frameworks for validation, verification, and sensitivity analysis (a recuring problem in biomechanical modelling research), and in particular, there are challenges in accurate soft tissue segmentation and reproduction, which is known to influence biomechanical simulation accuracy [11–14].

The present paper examined the behaviour of the spine motion segment in a hierarchical approach. The 3DP samples were produced, tested, and modelled at the material sample level (cores) at the vertebral body level, IVD level and at the analogue motion segment level. Validation took place first on the components and was then scaled up to a motion segment that combines all. Only one complete motion segment was modelled, and that may be seen as a limitation, but the subcomponents were created after scanning a larger number of vertebral samples as examples and testing multiple of the lower scale level components. The rationale was to establish the biofidelic accuracy of the 3DP process up and down the length scales, because the complexity exists in the subcomponents and their modelling, rather than in the testing of a whole analogue motion segment. We believe that the trade-off paid off, as the final complete AMS behaved very similarly to the in silico model. This was only possible through fine-tuning the lower length scale parts.

The creation of physical models of natural tissue structures by 3D printing and their computational equivalents are fraught with challenges at each length scale, because natural structures are hierarchically built and architecturally very complex. We have shown that it is possible to produce biofidelic spine motion segments in both analogue and digital (in silico) forms. There is still some distance to cover with regard to modelling spinal motion segments, particularly the intervertebral discs. There are two important aspects in producing physical analogues, the material level properties (IVD is an example in question)

https://doi.org/10.3390/jmmp10020056

and the length scale issues in scanning and 3D printing. The scanning equipment in this study uses a higher resolution than the current clinical CT scans, and to achieve this resolution in practice would require the exposure of patients to higher levels of radiation than is currently advisable. There is promising new research into CT dose reduction, which theoretically allows for the retrieval of much higher spatial resolutions in vivo without harm to patients [37]. The lack of a clear framework for validation, verification and sensitivity analysis [38] makes the case for a real need for further studies into the definition of accurate replicas in conjunction with FEA models and in paying particular attention to the segmentation of soft tissues in the spine, because this has shown to have a substantial effect on the biomechanical response of the analogues.

The creation of an accurate analogue means that tests involving many different procedures and devices will become more compatible and feasible due to the lack of intra-specimen variability and biohazard risks, thus making rapid prototyping ultimately an attractive alternative for in vitro spine biomechanical testing. Moreover, in cases where the biomechanical response of the spine is of paramount importance, the synthetic spine and its associated methods presented here have been shown to exhibit a reliable force/displacement response in compression loading at least. At this point, it must be mentioned that a further limitation of the present application was that the finer dimension of the trabecular elements of the artificial vertebral bodies were cruder (thicker) than the actual scanned trabeculae. This was a manufacturing limitation, in this instance, resulting in higher BV/TV ~0.8 values to the scanned bone samples in order to achieve the same loading response. In this respect, the current model remains a geometric approximation, rather than a functional biological analogue, necessitating further experimental validation and design to address bone-like physiological structures and loading conditions.

5. Conclusions

In conclusion, our present study has shown that it is practically feasible to produce near-biofidelic spine motion segments in two forms (modelling and physical), offering a more holistic approach for future work in the field. This fact advocates for greater use of synthetic models as a valid alternative for in vitro testing, because they offer standardisation, reproducibility and safety (no biohazard risks). For the design and production of bespoke and patient specific implants [39], the study supports the notion that rapid prototyping can be used to produce analogues for testing medical devices and surgical procedures, thus reducing the need for cadaveric and animal models. On clinical translation, it offers an example of a pathway which can integrate technologies such as computing and clinical scanning [2,5,6,23], which, with advances in radiation dose reduction and increasing resolution, will offer new avenues for applications.

Author Contributions: Conceptualisation: C.F. and P.Z.; Data curation: C.F. and T.S.; Formal analysis: C.F., T.S. and M.C.G.; Funding acquisition: C.F. and P.Z.; Investigation: C.F. and T.S.; Methodology: C.F.; Project administration: P.Z.; Resources: C.F.; Software: C.F., T.S. and M.C.G.; Supervision: P.Z.; Validation: C.F. and T.S.; Visualisation: C.F., T.S. and M.C.G.; Writing (original draft): C.F.; Writing (review and editing): C.F., T.S., M.C.G. and P.Z. All authors have read and agreed to the published version of the manuscript.

Funding: The authors disclosed receipt of the following financial support for the research, authorship, and/or publication of this article from a CDS Cranfield University internal funding to C.F. and T.S. and funding to P.Z. from RCDM (Royal College for Defence Medicine), Birmingham (grant no. DMSRSG20130905).

https://doi.org/10.3390/jmmp10020056

Institutional Review Board Statement: Ethical approval was granted by the Cranfield University Research and Ethics committee (CURES/2539/2017, Bone mechanobiology research studies group: spine strength assessment). Animal tissue from the food chain was used for the scans. No human tissue or data was used at any stage of this work.

Data Availability Statement: Data for this manuscript will be made available through the Cranfield University CORD data depository and preservation system (https://cranfield.figshare.com, accessed on 30 November 2025).

Acknowledgments: The main bulk of the material of this paper originates from the PhD Thesis of the first author (Chapter 6), which is listed in the Reference [19]. This paper is dedicated to one of the authors, Mike Gibson, whose untimely death was a great loss to us all.

Conflicts of Interest: The authors declare that the research was conducted in the absence of any commercial or financial relationships that could be construed as a potential conflict of interest.

Declaration AI: The authors categorically declare that AI has not been used anywhere in this work, either in conceptualisation, in carrying out the work, or in reporting.

Abbreviations

The following abbreviations are used in this manuscript:

3DP	3D printing
ABS	acrylonitrile butadiene styrene
AM	additive manufacturing
AMS	analogue motion segment
AVC	artificial vertebral cores
AVB	artificial vertebral body
BV/TV	bone volume/tissue volume
DA	degree of anisotropy
FDM	fused deposition modelling
IVD	intervertebral disc
LTC	cores with longitudinal and transverse pores artificially designed
FLC	FE model cores of idealised/artificially designed porosity architecture
F-DLS	FE model cores with internal trabecular architecture from scans

References

1. Agarwal, R.; Malhotra, S.; Gupta, V.; Jain, V. The application of Three-dimensional printing on foot fractures and deformities: A mini-review. *Ann. 3D Print. Med.* **2022**, *5*, 100046. [CrossRef]
2. Ventola, C.L. Medical Applications for 3D Printing: Current and Projected Uses. *Pharm. Ther.* **2014**, *39*, 704–711. [PubMed]
3. Burgade, R.; Uhl, J.F.; Delmas, V.; Dubois de Mont-Marin, G.; Rosset, P.; Le-Nail, L.R. Use of 3D printing as a simulation tool for trauma surgery of the pelvis. *Ann. 3D Print. Med.* **2021**, *3*, 100025. [CrossRef]
4. Biglino, G.; Capelli, C.; Wray, J.; Schievano, S.; Leaver, L.K.; Khambadkone, S.; Giardini, A.; Derrick, G.; Jones, A.; Taylor, A.M. 3D-manufactured patient-specific models of congenital heart defects for communication in clinical practice: Feasibility and acceptability. *BMJ Open* **2015**, *5*, e007165. [CrossRef] [PubMed]
5. Famaey, N.; Vander Sloten, J. Soft tissue modelling for applications in virtual surgery and surgical robotics. *Comput. Methods Biomech. Biomed. Eng.* **2008**, *11*, 351–366. [CrossRef]
6. Żukowska, M.; Rad, M.A.; Górski, F. Additive Manufacturing of 3D Anatomical Models-Review of Processes, Materials and Applications. *Materials* **2023**, *16*, 880. [CrossRef] [PubMed]
7. Jani, G.; Johnson, A.; Marques, J.; Franco, A. Three-dimensional(3D) printing in forensic science–An emerging technology in India. *Ann. 3D Print. Med.* **2021**, *1*, 100006. [CrossRef]
8. Biggs, M. 3D Printing Applied to Forensic Investigations. In *Essentials of Autopsy Practice*; Rutty, G., Ed.; Springer: Cham, Switzerland, 2019. [CrossRef]
9. Grassi, L.; Väänänen, S.P.; Yavari, S.A.; Weinans, H.; Jurvelin, J.S.; Zadpoor, A.A.; Isaksson, H. Experimental validation of finite element model for proximal composite femur using optical measurements. *J. Mech. Behav. Biomed. Mater.* **2013**, *21*, 86–94. [CrossRef] [PubMed]

 https://doi.org/10.3390/jmmp10020056

10. McDonald, K.; Little, J.; Pearcy, M.; Adam, C. Development of a multi-scale finite element model of the osteoporotic lumbar vertebral body for the investigation of apparent level vertebra mechanics and micro-level trabecular mechanics. *Med. Eng. Phys.* **2010**, *32*, 653–661. [CrossRef]
11. Dall'Ara, E.; Karl, C.; Mazza, G.; Franzoso, G.; Vena, P.; Pretterklieber, M.; Pahr, D.; Zysset, P. Tissue properties of the human vertebral body sub-structures evaluated by means of microindentation. *J. Mech. Behav. Biomed. Mater.* **2013**, *25*, 23–32. [CrossRef] [PubMed]
12. Helgason, B.; Gilchrist, S.; Ariza, O.; Vogt, P.; Enns-Bray, W.; Widmer, R.P.; Fitze, T.; Pálsson, H.; Pauchard, Y.; Guy, P.; et al. The influence of the modulus-density relationship and the material mapping method on the simulated mechanical response of the proximal femur in side-ways fall loading configuration. *Med. Eng. Phys.* **2016**, *38*, 679–689. [CrossRef]
13. Lee, C.-H.; Landham, P.R.; Eastell, R.; Adams, M.A.; Dolan, P.; Yang, L. Development and validation of a subject-specific finite element model of the functional spinal unit to predict vertebral strength. *Proc. Inst. Mech. Eng. Part H J. Eng. Med.* **2017**, *231*, 821–830. [CrossRef]
14. Viceconti, M.; Hunter, P. The Virtual Physiological Human: Ten Years After. *Annu. Rev. Biomed. Eng.* **2016**, *18*, 103–123. [CrossRef] [PubMed]
15. Franceskides, C.; Leger, T.; Horsfall, I.; Tozzi, G.; Gibson, M.; Zioupos, P. Evaluation of bone excision effects on a human skull model—I: Mechanical testing and digital image correlation. *Proc. Inst. Mech. Eng. Part H J. Eng. Med.* **2020**, *234*, 337–345. [CrossRef]
16. Franceskides, C.; Gibson, M.; Zioupos, P. Evaluation of bone excision effects on a human skull model—II: Finite element analysis. *Proc. Inst. Mech. Eng. Part H J. Eng. Med.* **2020**, *234*, 346–355. [CrossRef]
17. Barr, J.R.; Wang, T.; Pelletire, M.H.; Walsh, W.R. Initial Experience with Synthetic Spinal Motion Segments: Biomechanical. In Proceedings of the 60th Ann Meet Orthopaedic Research Society, New Orleans, LA, USA, 15–18 March 2014.
18. Campbell, J.; Imsdahl, S.; Ching, R. Evaluation of a synthetic L2-L5 spine model for biomechanical testing. In Proceedings of the 56th Ann Meet ORS, Orthopaedic Research Society, New Orleans, LA, USA, 6–9 March 2010.
19. Franceskides, C. Subject-Specific Spinal Functional Model of Hard and Soft Tissues. Ph.D. Thesis, Cranfield University, Shrivenham, UK, 2018. Available online: https://dspace.lib.cranfield.ac.uk/server/api/core/bitstreams/4002324f-9f18-4012-b0ff-e785f15067bf/content (accessed on 30 November 2025).
20. Franceskides, C.; Arnold, E.; Horsfall, I.; Tozzi, G.; Gibson, M.C.; Zioupos, P. Spinal motion segments—I: Concept for a subject-specific analogue model. *J. Bionic Eng.* **2020**, *17*, 747–756. [CrossRef]
21. Franceskides, C.; Arnold, E.; Horsfall, I.; Tozzi, G.; Gibson, M.C.; Zioupos, P. Spinal motion segments—II: Tuning and optimisation for biofidelic performance. *J. Bionic Eng.* **2020**, *17*, 757–766. [CrossRef]
22. Hambli, R. Micro-CT finite element model and experimental validation of trabecular bone damage and fracture. *Bone* **2013**, *56*, 363–374. [CrossRef] [PubMed]
23. Buie, H.R.; Campbell, G.M.; Klinck, R.J.; MacNeil, J.A.; Boyd, S.K. Automatic segmentation of cortical and trabecular compartments based on a dual threshold technique for in vivo micro-CT bone analysis. *Bone* **2007**, *41*, 505–515. [CrossRef] [PubMed]
24. Adams, G.J.; Cook, R.B.; Hutchinson, J.R.; Zioupos, P. Microarchitecture and morphology of bone tissue over a wide range of BV/TV assessed by micro-computed tomography and three different threshold backgrounds. *Med. Eng. Phys.* **2022**, *106*, 103828. [CrossRef]
25. Schneider, C.; Rasband, W.; Eliceiri, K. NIH Image to ImageJ: 25 years of image analysis. *Nat. Methods* **2012**, *9*, 671–675. [CrossRef] [PubMed]
26. Doube, M.; Kłosowski, M.M.; Arganda-Carreras, I.; Cordelières, F.P.; Dougherty, R.P.; Jackson, J.S.; Schmid, B.; Hutchinson, J.R.; Shefelbine, S.J. BoneJ: Free and extensible bone image analysis in ImageJ. *Bone* **2010**, *47*, 1076–1079. [CrossRef]
27. Adams, G.J. Quality Factors at the Material and Structural Level That Affect the Toughness of Human Cancellous Bone. Ph.D. Thesis, Cranfield University, Shrivenham, UK, 2017. Available online: http://dspace.lib.cranfield.ac.uk/handle/1826/15952 (accessed on 30 November 2025).
28. Keaveny, T.M.; Hayes, W.C. A 20-year perspective on the mechanical properties of trabecular bone. *J. Biomech. Eng.* **1993**, *115*, 534–542. [CrossRef] [PubMed]
29. Rodríguez, J.F.; Thomas, J.P.; Renaud, J.E. Mechanical behavior of acrylonitrile butadiene styrene (ABS) fused deposition materials. Experimental investigation. *Rapid Prototyp.* **2001**, *7*, 148–158. [CrossRef]
30. Dreischarf, M.; Zander, T.; Shirazi-Adl, A.; Puttlitz, C.M.; Adam, C.J.; Chen, C.S.; Goel, V.K.; Kiapour, A.; Kim, Y.H.; Labus, K.M.; et al. Comparison of eight published static finite element models of the intact lumbar spine: Predictive power of models improves when combined together. *J. Biomech.* **2014**, *47*, 1757–1766. [CrossRef]
31. Kuo, C.S.; Hu, H.T.; Lin, R.M.; Huang, K.H.Y.; Lin, P.C.; Zhong, Z.C.; Hseih, M.L. Biomechanical analysis of the lumbar spine on facet joint force and intradiscal pressure—A finite element study. *BMC Musculoskelet. Disord.* **2010**, *11*, 151–163. [CrossRef]
32. Burghardt, A.J.; Link, T.M.; Majumdar, S. High-resolution computed tomography for clinical imaging of bone microarchitecture. *Clin. Orthop. Relat. Res.* **2011**, *469*, 2179–2193. [CrossRef] [PubMed]

33. Zapata-Cornelio, F.Y.; Day, G.A.; Coe, R.H.; Sikora, S.N.F.; Wijayathunga, V.N.; Tarsuslugil, S.M.; Mengoni, M.; Wilcox, R.K. Methodology to Produce Specimen-Specific Models of Vertebrae: Application to Different Species. *Ann. Biomed. Eng.* **2017**, *45*, 2451–2460. [CrossRef]
34. Laffosse, J.M.; Odent, T.; Accadbled, F.; Cachon, T.; Kinkpe, C.; Viguier, E.; Sales de Gauzy, J.; Swider, P. Micro-computed tomography evaluation of vertebral end-plate trabecular bone changes in a porcine asymmetric vertebral tether. *J. Orthop. Res. Off. Publ. Orthop. Res. Soc.* **2010**, *28*, 232–240. [CrossRef]
35. Zander, T.; Dreischarf, M.; Timm, A.K.; Baumann, W.W.; Schmidt, H. Impact of material and morphological parameters on the mechanical response of the lumbar spine—A finite element sensitivity study. *J. Biomech.* **2017**, *28*, 185–190. [CrossRef]
36. Cai, Y.; Zioupos, P.; Márquez-Grant, N.; Budair, B.; Junaid, S. Producing micro-finite element models from real-time clinical CT scanners: Calibration, validation and material mapping strategies. *Front. Bioeng. Biotechnol.* **2025**, *13*, 1670428. [CrossRef]
37. Kitchen, M.J.; Buckley, G.A.; Gureyev, T.E.; Wallace, M.J.; Andres-Thio, N.; Uesugi, K.; Yagi, N.; Hooper, S.B. CT dose reduction factors in the thousands using X-ray phase contrast. *Nat. Sci. Rep.* **2017**, *7*, 15953. [CrossRef] [PubMed]
38. Jones, A.C.; Wilcox, R.K. Finite element analysis of the spine: Towards a framework of verification, validation and sensitivity analysis. *Med. Eng. Phys.* **2008**, *30*, 1287–1304. [CrossRef] [PubMed]
39. Zhang, G.Q.; Li, J.X.; Li, J.; Zhou, X.Y.; Xie, J.J.; Wang, A.M. Selective Laser Melting Molding of Individualized Femur Implant: Design, Process, Optimization. *J. Bionic Eng.* **2021**, *18*, 128–137. [CrossRef]

https://doi.org/10.3390/jmmp10020056

Journal of
Manufacturing and Materials Processing

MDPI

Article

Design and Material Characterisation of Additively Manufactured Polymer Scaffolds for Medical Devices

Aidan Pereira [1], Amirpasha Moetazedian [2], Martin J. Taylor [2], Frances E. Longbottom [3,4], Heba Ghazal [5], Jie Han [6] and Bin Zhang [1,*]

1 Department of Mechanical and Aerospace Engineering, Brunel University of London, London UB8 3PH, UK; aidangarethpereira@gmail.com
2 School of Engineering and Technology, University of Hull, Hull HU6 7RU, UK; a.moetazedian@hull.ac.uk (A.M.); martin.taylor@hull.ac.uk (M.J.T.)
3 School of Natural Science, Chemistry, University of Hull, Hull HU6 7RX, UK; f.longbottom-2018@hull.ac.uk
4 Energy and Environment Institute, University of Hull, Hull HU6 7RX, UK
5 Department of Pharmacy, Kingston University London, London KT1 2EE, UK; h.ghazal@kingston.ac.uk
6 Department of Life Sciences, Brunel University of London, London UB8 3PH, UK; jie.han@brunel.ac.uk
* Correspondence: bin.zhang@brunel.ac.uk

Abstract

Additive manufacturing has been adopted in several industries including the medical field to develop new personalised medical implants including tissue engineering scaffolds. Custom patient-specific scaffolds can be additively manufactured to speed up the wound healing process. The aim of this study was to design, fabricate, and evaluate a range of materials and scaffold architectures for 3D-printed wound dressings intended for soft tissue applications, such as skin repair. Multiple biocompatible polymers, including polylactic acid (PLA), polyvinyl alcohol (PVA), butenediol vinyl alcohol copolymer (BVOH), and polycaprolactone (PCL), were fabricated using a material extrusion additive manufacturing technique. Eight scaffolds, five with circular designs (knee meniscus angled (KMA), knee meniscus stacked (KMS), circle dense centre (CDC), circle dense edge (CDE), and circle no gradient (CNG)), and three square scaffolds (square dense centre (SDC), square dense edge (SDE), and square no gradient (SNG), with varying pore widths and gradient distributions) were designed using an open-source custom toolpath generator to enable precise control over scaffold architecture. An in vitro degradation study in phosphate-buffered saline demonstrated that PLA exhibited the greatest material stability, indicating minimal degradation under the tested conditions. In comparison, PVA showed improved performance relative to BVOH, as it was capable of absorbing a greater volume of exudate fluid and remained structurally intact for a longer duration, requiring up to 60 min to fully dissolve. Tensile testing of PLA scaffolds further revealed that designs with increased porosity towards the centre exhibited superior mechanical performance. The strongest scaffold design exhibited a Young's modulus of 1060.67 ± 16.22 MPa and withstood a maximum tensile stress of 21.89 ± 0.81 MPa before fracture, while maintaining a porosity of approximately 52.37%. This demonstrates a favourable balance between mechanical strength and porosity that mimics key properties of engineered tissues such as the meniscus. Overall, these findings highlight the potential of 3D-printed, patient-specific scaffolds to enhance the effectiveness and customisation of tissue engineering treatments, such as meniscus repair, offering a promising approach for next-generation regenerative applications.

Keywords: 3D printing; material extrusion; tensile property; degradation; PLA; PVA; BVOH; PCL

1. Introduction

The musculoskeletal system contains several load-bearing joint tissues, such as cartilage, ligaments, and the meniscus, that play a critical role in joint stability, shock absorption, and load distribution [1–3]. The meniscus is a fibrocartilaginous tissue that experiences complex multiaxial loading during joint motion, including compressive, shear, and tensile stresses [4,5]. In particular, tensile stresses, generated as compressive loads, are converted into circumferential tension, enabling the meniscus to distribute loads across the knee joint effectively [6,7]. Damage to the meniscus disrupts this stress distribution mechanism, leading to increased contact stresses on the articular cartilage and an elevated risk of joint degeneration and osteoarthritis [8,9].

Due to its limited vascularisation, especially in the inner avascular region, the meniscus exhibits a poor intrinsic healing capacity, making effective repair challenging [10,11]. Current clinical treatments, including partial meniscectomy and grafting, are associated with limitations such as altered mechanical behaviour, incomplete tissue regeneration, and long-term deterioration of joint function. As a result, tissue engineering approaches that employ customised scaffolds have gained increasing interest for meniscus repair [12,13]. Such scaffolds must not only replicate the anatomical geometry of the native tissue but also provide sufficient tensile strength and controlled porosity to withstand physiological loading while supporting cell infiltration and tissue regeneration [14,15].

Material extrusion additive manufacturing (MEAM) has received substantial attention and investment in high-value sectors, including the medical industry, due to its versatility, accessibility, and compatibility with a wide range of thermoplastic polymers [16]. MEAM operates by sequentially depositing molten polymer layers onto a build platform; filament is fed into a heated chamber above the polymer's melting temperature and extruded through a nozzle onto the build surface [17]. Scaffold geometry is defined by the controlled movement of the nozzle in the x–y plane, combined with incremental displacement along the z-axis to form three-dimensional structures. Recent advances in custom toolpath generation [17] and four-axis printing [18] have significantly expanded the design freedom of MEAM, enabling the fabrication of complex structures that were previously unattainable using conventional CAD–slicer–print workflows. These developments allow the creation of scaffolds that conform to irregular wound topographies and support non-planar printing strategies. Moreover, MEAM enables the deposition of multiple materials within a single construct, including antibacterial agents or bioactive components, making it particularly attractive for personalised and functional wound dressing applications [19–21]. Currently, several biocompatible and bioresorbable polymers, including polylactic acid (PLA), polyglycolic acid (PGA), and polycaprolactone (PCL), are widely employed in MEAM owing to their controlled degradation behaviour and favourable printability.

A recent review by Uchida et al. [14] highlighted the advantages of MEAM-based scaffolds in promoting wound healing and tissue regeneration compared to conventional "one-size-fits-all" dressings. However, most existing studies [22–24] are limited to relatively simple scaffold architectures due to constraints imposed by traditional slicing software. In contrast, only a limited number of studies [17,25] have explored the use of custom toolpath generation to precisely control nozzle motion and scaffold architecture. Direct control of the printing toolpath enables the fabrication of intricate, highly porous, and open scaffold structures that do not completely seal the wound surface. These features allow healthcare professionals to visually inspect and clean the wound while enabling excess exudate produced during the inflammatory phase of healing to be absorbed through the scaffold structure. In this context, the scaffold can function as a semi-permeable membrane,

https://doi.org/10.3390/jmmp10010039

facilitating fluid management while protecting the wound from mechanical damage during handling or dressing changes [26,27]. To the best of our knowledge, very few previous studies have systematically investigated the combined effects of scaffold architecture, porosity distribution, mechanical performance, and degradation behaviour of MEAM-fabricated tissue engineering meniscus scaffolds using a custom toolpath generation approach across multiple biocompatible polymers.

In this study, PLA, PVA, BVOH, and PCL were selected to evaluate their suitability for personalised meniscus tissue engineering applications. PLA and PCL were chosen as primary structural materials due to their established use in load-bearing biomedical scaffolds, where mechanical strength, controlled degradation, and biocompatibility are essential. PVA and BVOH were included as water-soluble, bioresorbable polymers to investigate their potential roles as temporary or sacrificial materials in meniscus repair strategies. Their dissolution behaviour enables controlled removal or porosity evolution without mechanical intervention, while allowing comparison of degradation behaviour for applications requiring transient structural support.

2. Materials and Methods

2.1. Materials

Several polymeric materials were employed for the fabrication of three-dimensional (3D) printed scaffolds, selected based on their biocompatibility, biodegradability, and suitability for extrusion-based additive manufacturing. These included polylactic acid (PLA, Verbatim GmbH, Eschborn, Germany), polyvinyl alcohol (PVA, Bambu labs, Shenzhen, China), butenediol vinyl alcohol copolymer (BVOH, Verbatim GmbH, Eschborn, Germany), and polycaprolactone filament (PCL, 3D4Makers Ltd., Haarlem, The Netherlands). More details regarding filament-related processing and material parameters are shown in Table 1. For hygroscopic filaments such as BVOH and PVA, appropriate control of printing and handling conditions is essential. These filaments were dried prior to printing and processed using sealed containers. After printing, samples were cooled and stored immediately in airtight containers with desiccant to minimise exposure to ambient air. All materials were processed using extrusion-based 3D printing techniques under optimised printing parameters to ensure consistent filament deposition, pore uniformity, and structural integrity of the scaffolds.

Table 1. Filament-related processing and material parameters *.

Filament	Specific Gravity (g/cm^3)	Heat Distortion Temperature at 0.45 MPa (°C)	Glass Transition Temperature (°C)	Tensile Strength (MPa)	Tensile Young's Modulus (MPa)	Cost (GBP/g) (exc. VAT)	Ref.
PLA	1.31	52	60.60	20.9 ± 2.0	1882 ± 141	0.03	[28,29]
BVOH	1.14	63	68	34	2300	0.09	[30–32]
PVA	1.1	66.5	70	35.70 ± 1.70	1745 ± 151	0.06	[33]
PCL	1.1	57	−60	45	350	0.10	[34]

* For certain filaments, manufacturers provide limited data due to formulation confidentiality or moisture sensitivity, and additional filament-related data are obtained from relevant references.

2.2. Scaffold Design and Additive Manufacturing

A Prusa i3 Mk3S material extrusion additive manufacturing (MEAM) system (Prusa Ltd., Prague, Czech Republic) was utilised to fabricate complex scaffold structures with square and circular geometries (Figure 1). The scaffolds were generated using FullControl GCode Designer excel version [35], which enables direct manipulation of the printer toolpath. This approach provides precise control over critical printing parameters, includ-

 https://doi.org/10.3390/jmmp10010039

ing extrusion rate, printing speed, nozzle temperature, and acceleration, offering greater flexibility and accuracy compared to conventional slicer-based software as described previously [36]. The printing parameters applied in this study are summarised in Table 2.

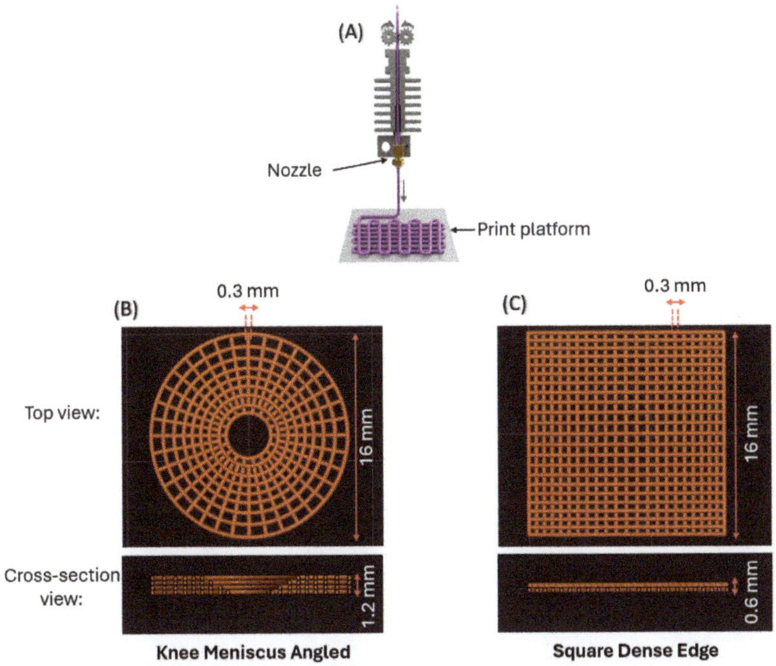

Figure 1. (**A**) The MEAM setup used to print the various scaffold designs. (**B**) Top and cross-sectional views of an angled knee meniscus scaffold comprising four layers. (**C**) Top and cross-sectional views of a square dense-edge scaffold comprising two layers. All designs were fabricated with a constant line width of 0.3 mm.

Table 2. Design parameters used to produce all scaffolds using Prusa MEAM system.

Design	Diameter or Length (mm)	Line Width (mm)	Line Height (mm)	Number of Layers	Smallest Pore Width (mm)	Largest Pore Width (mm)	Gradient
KMA	Ø16	0.300	0.300	4	0.385	2.310	Denser at the Centre
KMS	Ø16	0.300	0.300	4	0.385	2.310	Denser at the Centre
SDC	15.15	0.300	0.300	2	0.300	0.675	Denser at the Centre
SDE	16	0.300	0.300	2	0.300	0.575	Denser at the Edges
CDC	Ø16.5	0.300	0.300	2	0.300	0.600	Denser at the Centre
CDE	Ø16.5	0.300	0.300	2	0.300	1.500	Denser at the Edges
CNG	Ø18	0.300	0.300	2	0.300	0.300	No Gradient
SNG	16.8	0.300	0.300	2	0.300	0.300	No Gradient

The scaffold designs were created (see Table 2) and continuously improved upon until the ideal print parameters were identified. Onshape (Onshape Ltd., Boston, MA, USA), a CAD (Computer-Aided Design) programme, and Repetier Host (Hot-World GmbH & Co. KG, Willich, Germany) were used to visualise the design so that design dimensions and constraints could be adjusted without the need for 3D printing. This was enacted to reduce unnecessary wastage of filament. Onshape was used to create a CAD model of the design so that crucial dimensions could be extracted as this was found to be quicker than solving large trigonometric equations. It also helped with calculating expected pore sizes and preventing overlapping of filament fibres. Repetier Host was used to visualise the generated G-code to

 https://doi.org/10.3390/jmmp10010039

verify how the design is expected to appear. This was a useful tool as it gives a reference point for the design's appearance that can be used as a benchmark for the actual 3D prints. The design and print parameters for the final designs can be seen in Table 2. The designs were named based on their description, as follows: (1) KMA (knee meniscus angled): the knee meniscus scaffold design where each layer has a polar shift; (2) KMS (knee meniscus stacked): the knee meniscus scaffold design where each layer is perfectly stacked over each other (no polar shift); (3) SDC (square dense centre): square-shaped scaffold design where the porosity reduces towards the centre, i.e., the centre of the scaffold is denser; (4) SDE (square dense edge): square-shaped scaffold design where the porosity increases towards the centre, i.e., the edges of the scaffold are denser; (5) CDC (circle dense centre): circle-shaped scaffold design where the porosity reduces towards the centre, i.e., the centre of the scaffold is denser; (6) CDE (circle dense edge): circle-shaped scaffold design where the porosity increases towards the centre, i.e., the regions near the perimeter of the scaffold are denser; (7) CNG (circle no gradient): circle-shaped scaffold design where the porosity remains uniform, i.e., porosity remains constant towards the centre; (8) SNG (square no gradient): square-shaped scaffold design where the porosity remains uniform, i.e., porosity remains constant towards the centre.

2.3. Thermogravimetric Analysis (TGA)

Thermogravimetric analysis (TGA) was conducted using a LECO 701 TGA system (LECO, St. Joseph, MI, USA) to assess the thermal behaviour of the samples. The samples were heated from ambient temperature to 107 °C at 3 °C min^{-1} under nitrogen and held for 15 min to remove moisture. The temperature was then increased to 950 °C at 5 °C min^{-1} and held for 7 min under nitrogen to volatilize organic components. After cooling to 600 °C, the atmosphere was switched to air, and the samples were heated to 750 °C at 3 °C min^{-1} to oxidise residual material for the ashing phase. Due to the overlap of heating, the ashing phase is not illustrated on the analysed data.

2.4. Differential Scanning Calorimetry (DSC)

Differential scanning calorimetry (DSC) was performed using a PerkinElmer DSC 4000 (PerkinElmer, Shelton, CI, USA) to evaluate the thermal transitions and thermal stability of the materials. The measurements were carried out under a nitrogen atmosphere to prevent oxidative degradation. Samples were heated from 30 °C to 450 °C, corresponding to the temperature at which sample destruction occurs, at a constant heating rate of 5 °C min^{-1}. The DSC instrument was operated in parallel with an indium standard for temperature and enthalpy calibration to ensure measurement accuracy and reproducibility. The resulting thermograms were used to identify characteristic thermal activities including glass transition and melting behaviour of the plastic materials.

2.5. Powder X-Ray Diffraction (PXRD)

Powder X-ray diffraction (PXRD) measurements were performed using a PANalytical Empyrean Series 2 diffractometer (Malvern Panalytical, Malvern, UK) with monochromated Cu Kα radiation (λ = 0.1542 nm). Diffraction patterns were analysed using HighScore Plus software (version 2013, PANalytical B.V., Malvern Panalytical, Malvern, UK), and phase identification was carried out by comparison with reference data from the ICDD PDF-2 database (2026 release) to assess the materials for crystallinity and/or elemental additions.

2.6. Scanning Electron Microscopy (SEM)

The surface morphology of the 3D-printed samples was examined using scanning electron microscopy (SEM) with a Zeiss Gemini 300 system (Jeol IT200LV, JEOL, Peabody, MA, USA). Prior to imaging, the samples were sputter-coated with a thin layer of gold (~10 nm) to enhance surface conductivity, increase contrast and minimise charging ef-

https://doi.org/10.3390/jmmp10010039

fects during observation. SEM images were acquired over a range of magnifications from ×25 to ×150, using an accelerating voltage of 10 kV. This analysis enabled detailed evaluation of surface features, filament deposition, pore structure, and overall scaffold morphology.

2.7. Mechanical Tensile Test

To investigate the influence of scaffold design on the mechanical properties of 3D-printed parts, PLA was chosen as the primary material for mechanical testing. Other materials, including PVA, BVOH, and PCL, were excluded from tensile testing due to their susceptibility to moisture absorption and subsequent degradation, which could affect the accuracy and reliability of the mechanical properties.

PLA scaffolds were subjected to tensile testing using a TA.XTplusC texture analyser (Stable Micro Systems, Godalming, UK) to assess their mechanical performance under stress. To ensure proper clamping of each scaffold during mechanical testing, all designs were scaled by a factor of two in the x- and y-directions. Despite this geometric scaling, the cross-sectional area perpendicular to the applied load was maintained at the same value as in the original designs by preserving key printing parameters, including line width and layer height. As a result, the effective load-bearing cross-section remained unchanged, ensuring that the measured mechanical properties were directly comparable to those of the unscaled designs.

The tensile tests were conducted by stretching each sample until reaching 20% strain, as described in the study by Ottenio et al. [37]. Initially, the texture analyser recorded the force (N) versus distance (mm) data. This data was subsequently transformed into stress versus strain curves, allowing for the calculation of mechanical properties, such as Young's modulus, yield strength, and ultimate tensile strength. For each scaffold design, three samples (n = 3) were tested, and the mechanical properties were averaged to ensure statistical reliability.

The applied force was subsequently converted to engineering stress by dividing the measured force by the cross-sectional area of each scaffold, as shown in Equation (1). This normalisation accounts for differences in specimen geometry and enables direct comparison of the mechanical response between different scaffold designs.

$$Stress\ (\mathrm{MPa}) = Force\ (\mathrm{N}) / Cross\ sectional\ Area\ \left(\mathrm{mm}^2\right) \quad (1)$$

Due to the shape of the printer's nozzle and the print parameters, namely the line width and layer height, both being 0.3 mm, the print is created by depositing material along the path in the shape that can be approximated to that of a circle of diameter 0.3 mm. Thus, the cross-section can be seen in Figure 1B,C and the area was calculated in CAD by cutting cross-sections of each scaffold. The square scaffold comprises two layers, whereas the KMA and KMS designs incorporate two additional layers to form a hemispherical cavity that accommodates the femoral condyle, thereby mimicking the geometry of the knee meniscus, as shown in Figure 1B. The total cross-sectional area of the scaffold was then obtained by summing the calculated areas across all print layers. The resulting cross-sectional areas corresponding to each scaffold design are presented in Table 3.

Strain was calculated by dividing the measured displacement by the original gauge length or diameter of each scaffold design, as shown in Equation (2). This approach provides a normalised measure of deformation, allowing direct comparison of the mechanical response across different scaffold geometries.

$$Strain\ (\varepsilon) = Displacement\ (\mathrm{mm}) / Original\ Length\ or\ Diameter\ (\mathrm{mm}) \quad (2)$$

https://doi.org/10.3390/jmmp10010039

Young's modulus was determined by plotting stress as a function of strain and calculating the slope of the elastic region, which corresponds to the initial linear portion of the stress–strain curve. The elastic region was identified using the 0.2% strain offset method, which allows for consistent determination of elastic behaviour in materials that do not exhibit a well-defined yield point. The intersection between the stress–strain curve and the 0.2% offset line was taken as the yield point of the scaffold structure. In addition to Young's modulus and yield strength, other key mechanical properties, including the ultimate tensile strength (UTS), were obtained directly from the stress–strain curves.

Table 3. Cross-sectional area corresponding to each scaffold design, calculated perpendicular to the direction of the applied tensile load.

Design	Cross-Sectional Area (mm^2) for Layer No.				Total Cross-Sectional Area (mm^2)
	1	2	3	4	
KMA	3.81	3.45	2.97	2.55	12.78
KMS	4.11	3.39	2.97	2.55	13.02
SDC	1.98	1.98	N/A	N/A	3.96
SDE	2.07	2.07	N/A	N/A	4.14
CDC	1.98	1.98	N/A	N/A	3.96
CDE	2.07	2.07	N/A	N/A	4.14
CNG	2.61	2.61	N/A	N/A	5.22
SNG	2.52	2.52	N/A	N/A	5.04

2.8. Calculating Porosity

The porosity of the printed scaffold structures was determined by comparing their measured mass to that of a hypothetical solid structure with identical external geometry and dimensions. This mass-based approach enables estimation of the internal void fraction within each scaffold design.

To calculate the mass of the solid-equivalent structure, the density of the printing filament was first determined experimentally. Sections of filament were cut into three separate samples, and their mass and geometric dimensions were measured. The filament density (Equation (3)) was calculated using the average mass and volume of the three samples. The volume of each filament sample (Equation (4)) was determined by multiplying the cross-sectional area by its height, which in this case corresponds to the filament length. Using the calculated filament density, the mass of a fully solid structure with the same external dimensions as the scaffold was then estimated.

$$Filament\ Density = \frac{Filament\ Mass}{Filament\ Volume} \tag{3}$$

$$Volume = \pi \times Radius^2 \times Height \tag{4}$$

Porosity was subsequently calculated by comparing the measured mass of each printed scaffold with the estimated mass of its fully solid (0% porosity) equivalent. The difference between these two values represents the volume fraction of pores within the structure, allowing porosity to be expressed as a percentage. This mass-based approach provides a consistent and reproducible method for comparing porosity across different scaffold designs. Once the density of the filament had been determined, the mass of the solid (0% porosity) equivalent structure could be calculated by rearranging Equation (3) to get Equation (5), as shown below:

$$Scaffold\ Mass = Filament\ Density \times Scaffold\ Volume \tag{5}$$

https://doi.org/10.3390/jmmp10010039

Equation (4) was used to calculate the volume of the cylindrical scaffold designs (KMA, KMS, CDC, CDE and CNG), while Equation (6) was applied to determine the volume of the cuboidal scaffold designs (SDC, SDE and SNG). These equations were selected based on the external geometry of each scaffold to ensure accurate volume estimation for subsequent porosity calculations. For each scaffold design, three samples were analysed, and the average volume and porosity values were calculated to improve measurement reliability and reproducibility.

$$Scaffold\ Volume = Length \times Width \times Height \tag{6}$$

$$Porosity\ (\%) = 100 \times \left(1 - \frac{Mass\ of\ scaffold}{Mass\ of\ Equivalent\ Solid\ Body}\right) \tag{7}$$

Equation (7) was then used to calculate the volumetric porosity of the scaffold designs. In addition, an optical microscope (AmScope, Irvine, CA, USA) was employed during the initial stages of fabrication to examine the resulting pore sizes and verify printing accuracy. Based on these observations, printing parameters, including initial layer height and extrusion rate, were adjusted as necessary to ensure that the experimentally achieved pore sizes closely matched the intended design dimensions, within the tolerances of the 3D printer and the microscope.

2.9. In Vitro Degradation Test

To evaluate the in vitro degradation behaviour and stability of the scaffolds, degradation tests were conducted on PLA, BVOH, PCL and PVA samples. Three-dimensional printed specimens from each material were placed in individual containers containing 30 mL of 0.1 M phosphate-buffered saline (PBS) and incubated at 37 °C to simulate physiological conditions. The study was carried out for up to 60 min. Prior to immersion, samples were dried at room temperature at 21 °C and weighed to obtain their initial mass. At predetermined time intervals, the samples were removed, gently dried on paper tissue to remove surface moisture, and reweighed to monitor mass changes.

During testing, noticeable softening of the BVOH and PVA scaffolds made handling difficult. To address this, a customised perforated ladle was designed and 3D-printed to allow safe handling of partially degraded samples. After each measurement, the samples were returned to the oven to maintain a constant temperature of 37 °C.

Following completion of the degradation tests, the PBS solutions were maintained at 37 °C for an additional 48 h to ensure complete dissolution of soluble degradation products. The pH of each solution was then measured using a calibrated pH meter (Mettler Toledo Ltd., Columbus, OH, USA) and compared with fresh PBS to assess changes in solution acidity associated with scaffold degradation.

2.10. Statistical Analysis

Numerical data was calculated and reported as the mean ± standard deviation (SD), with each experimental condition tested in triplicate. Statistical analysis was performed using Student's t-test to evaluate differences between groups. A p-value ≤ 0.05 was considered statistically significant, whereas a p-value > 0.05 indicated no statistically significant difference between the groups.

3. Results and Discussion

3.1. Design Freedom Through Custom Toolpath Generation

By directly controlling the printer toolpath, rather than relying on a conventional CAD–slicer–print workflow, a range of architecturally complex scaffold structures were

https://doi.org/10.3390/jmmp10010039

fabricated using PLA with a constant line width and layer height of 0.3 mm (Figure 2). This approach highlights the versatility of the custom toolpath generation software in enabling precise control over scaffold geometry and porosity distribution.

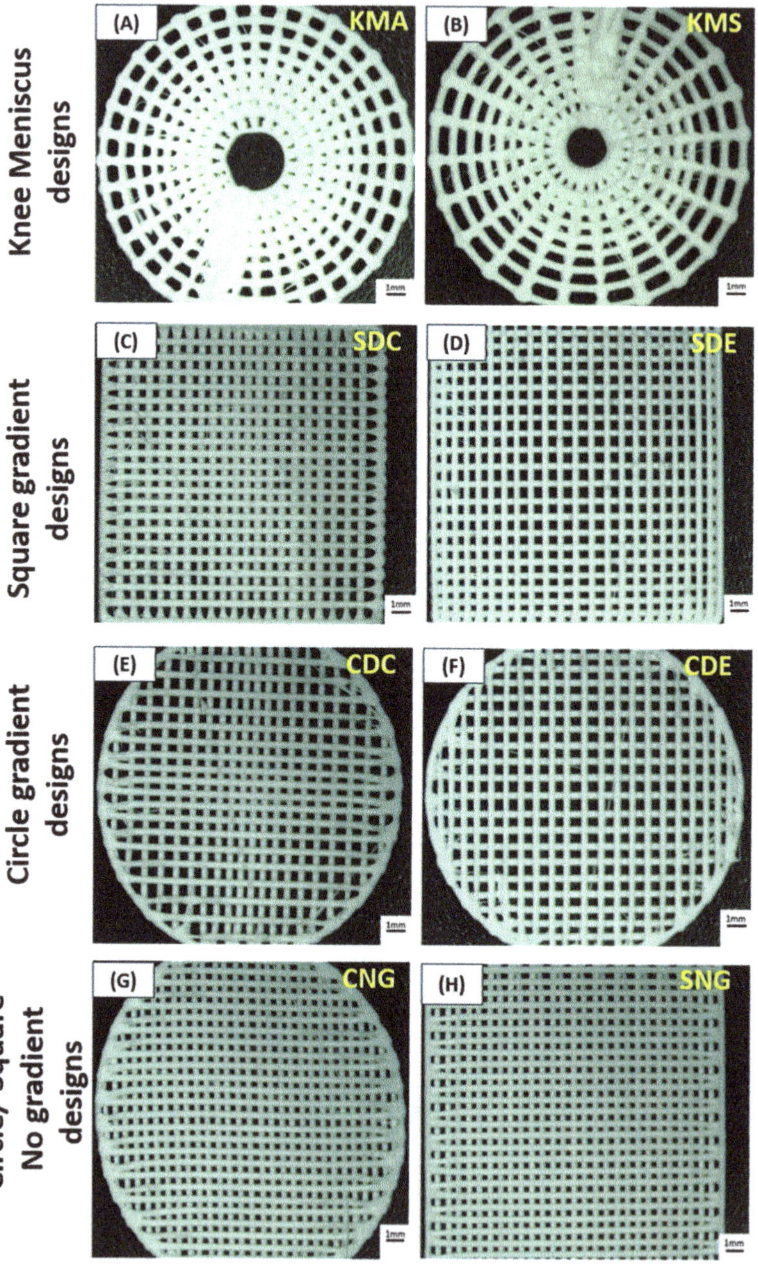

Figure 2. Optical images of the different PLA scaffold designs: (**A**) knee meniscus angled (KMA), (**B**) knee meniscus stacked (KMS), (**C**) square dense centre (SDC), (**D**) square dense edge (SDE), (**E**) circle dense centre (CDC), (**F**) circle dense edge (CDE), (**G**) circle no gradient (CNG) and (**H**) square no gradient (SNG). Scale bars represent 1 mm.

https://doi.org/10.3390/jmmp10010039

The knee meniscus angled (KMA) and knee meniscus stacked (KMS) designs (Figure 2A,B) each consist of four layers in which the internal void progressively expands with height, creating a vertical porosity gradient. In the KMA design, successive layers are rotated by 20°, introducing a polar offset between layers, whereas in the KMS design no inter-layer rotation is applied, resulting in vertically stacked fibres. This vertical gradient was incorporated to better mimic the curved internal cavity and load-distribution characteristics of the native meniscus.

The circle dense centre (CDC) and circle dense edge (CDE) designs (Figure 2E,F) comprise two identical layers of unidirectional lines, with the second layer rotated by 90°. Both designs incorporate a controlled radial density gradient, with CDC exhibiting increased density at the centre and CDE at the periphery. Similarly, the square dense centre (SDC) and square dense edge (SDE) designs (Figure 2C,D) employ two orthogonally oriented layers with graded infill density, where material concentration is either central or edge-dominant. In contrast, the circle no gradient (CNG) and square no gradient (SNG) designs (Figure 2G,H) consist of two orthogonal layers without any spatial density variation.

The minimum pore size for all gradient designs was set to 0.3 mm, while the maximum pore size was 1.2 mm. These values were selected based on previous findings by Saijo-Rabina et al. [38], which demonstrated that pore sizes within this range promote effective fluid transport through capillary action. For non-gradient designs, pore sizes were maintained at 0.3 mm, as smaller pores (approximately 400 µm) have been shown to provide improved performance in applications where controlled fluid management is required. To ensure consistency across all designs, printing parameters including line width and layer height were kept constant at 0.3 mm. This allowed the influence of scaffold architecture, porosity gradients, and layer orientation on mechanical and degradation behaviour to be evaluated independently of printing resolution effects.

3.2. Material Effects on Complex Structure Printing

To further assess the transferability of the proposed scaffold designs to other biocompatible polymers suitable for tissue engineering, all eight scaffold architectures listed in Table 2 were fabricated using PLA, BVOH, PVA, and PCL under the printing conditions summarised in Table 4. Optical images of the resulting constructs are shown in Figure 3. A qualitative comparison of the printed samples revealed a clear trend in print quality, which decreased in the following order: PLA > PCL > BVOH > PVA. This observation was consistent across all scaffold designs and was further supported by the surface morphology analysis discussed below.

Table 4. MEAM process parameters for the various materials used to fabricate the designed scaffold geometries.

Filament	Manufacturer-Recommended Nozzle Temp. (°C)	Nozzle Temp. (°C) (Experimental)	Manufacturer-Recommended Bed Temp. (°C)	Bed Temperature (°C) (Experimental)	Company
PLA	200–220	220	60	60	(Verbatim GmbH, Germany)
BVOH	200–220	220	60	60	(Verbatim GmbH, Germany)
PVA	220–250	230	35–45	40	(Bambu Labs, China)
PCL	130–170	190	30–45	0 (Room Temperature; bed heating turned off)	(3D4Makers B.V, Netherlands)

 https://doi.org/10.3390/jmmp10010039

Figure 4 presents SEM images of all four materials acquired at magnifications of ×25 and ×150, allowing direct comparison of surface morphology and roughness. PLA (Figure 4A,B) and PCL (Figure 4C,D) present relatively smooth and continuous surfaces with well-defined filament boundaries, indicating stable extrusion and consistent material flow during printing. In particular, PLA demonstrated the highest fidelity to the intended design geometry, with minimal surface defects. This superior printability can be attributed to its favourable melt viscosity, wide processing window, thermal stability, and low sensitivity to environmental moisture. Unlike the other materials, PLA did not require specialised storage conditions and remained relatively unaffected by ambient humidity. Optimal print quality for PLA was achieved at a nozzle temperature of 220 °C and a bed temperature of 60 °C, as summarised in Table 4.

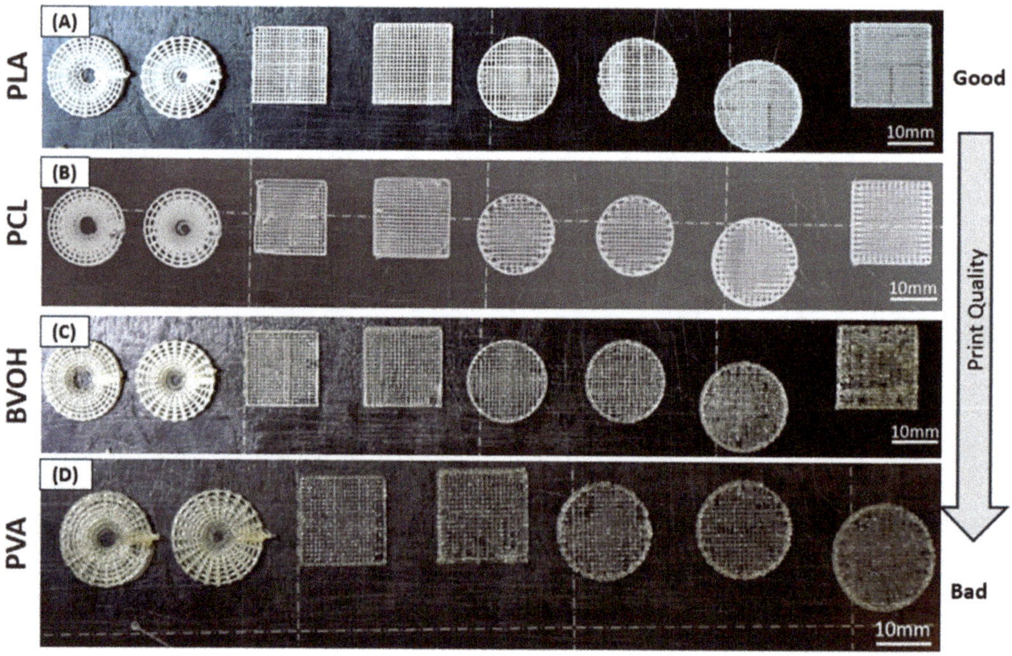

Figure 3. Effect of material properties on the 3D printing of different scaffold designs. All primary scaffold architectures were fabricated using (**A**) PLA, (**B**) PCL, (**C**) BVOH and (**D**) PVA. PLA exhibited the highest print quality, while PVA showed multiple printing defects. Scale bars represent 10 mm.

In contrast, BVOH (Figure 4E,F) and PVA (Figure 4G,H) displayed noticeably rougher surfaces with pits, voids and irregular filament deposition. These defects are largely attributed to the hygroscopic nature of both polymers, which readily absorb moisture from the environment, evidenced by a minor weight loss for each polymer when heating to 107 °C (Figure 5A–D). Despite storage in sealed containers with desiccant and using filament dryers, both materials remained prone to moisture-related printing issues. During extrusion, absorbed moisture can vaporise, leading to bubble formation, surface pitting, and inconsistent material flow. In addition, the relatively low mechanical strength of BVOH and PVA filaments resulted in frequent filament snapping within the extruder, often necessitating partial disassembly of the extruder and hotend to remove lodged filament. BVOH printed optimally at the same temperature settings as PLA, allowing the same G-code to be reused without modification, whereas PVA required a higher nozzle temperature of 230 °C and a lower bed temperature of 40 °C (Table 4).

 https://doi.org/10.3390/jmmp10010039

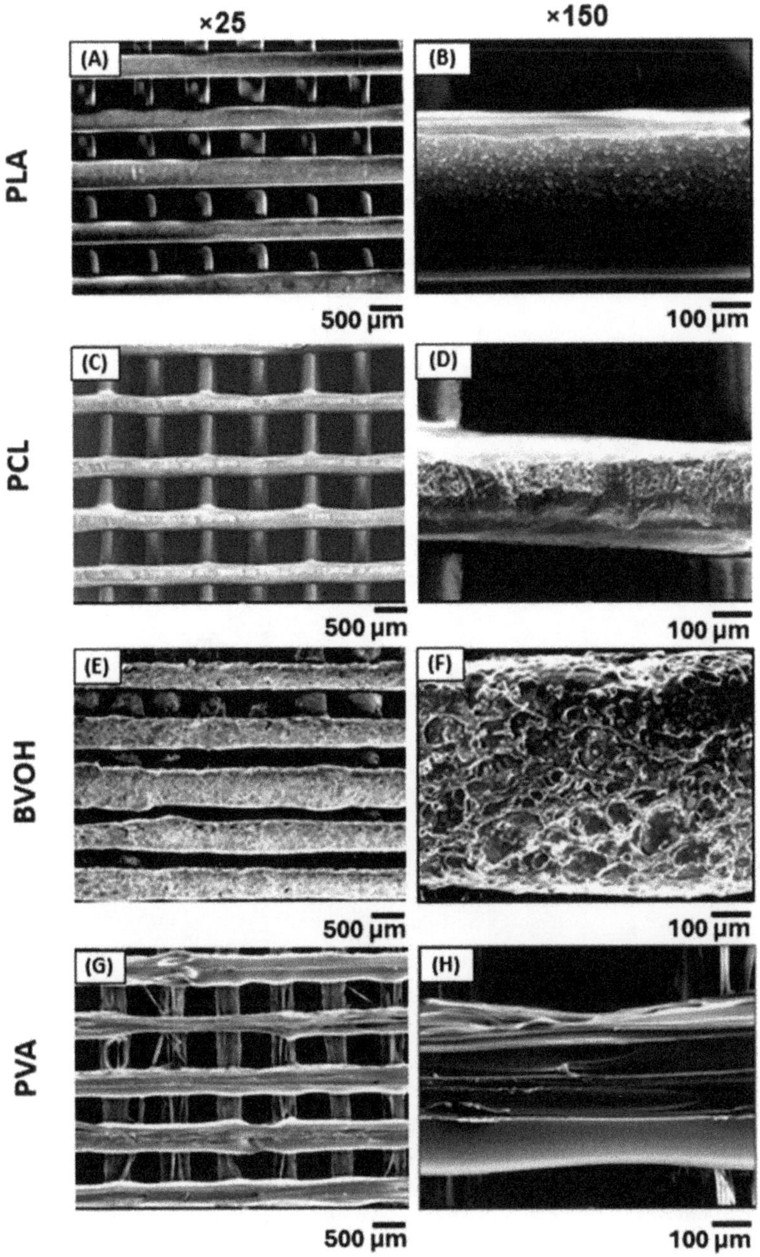

Figure 4. SEM micrographs of 3D-printed scaffolds acquired at two different magnifications. (**A**,**B**) PLA and (**C**,**D**) PCL exhibited superior print quality with smoother and more uniform surfaces compared to (**E**,**F**) BVOH and (**G**,**H**) PVA, which showed increased surface roughness and printing defects.

PCL exhibited intermediate surface quality. The filament was noticeably softer than the other polymers and frequently deformed or fractured within the extruder mechanism. Successful printing required nozzle and bed temperature settings outside those recommended by the filament manufacturer (3D4Makers, Haarlem, The Netherlands). Optimal

https://doi.org/10.3390/jmmp10010039

extrusion was achieved at a nozzle temperature of 190 °C with the print bed unheated (Table 4), as PCL did not melt sufficiently at the manufacturer-specified temperatures. Reducing the tension on the extruder drive gears helped to mitigate filament breakage, although extrusion failures still remained relatively common. It is worthwhile to mention that the primary difference between the G-code used for each material was limited to the start code, which defines the temperature settings for the nozzle and print bed. As all filaments shared identical nominal diameters, the remainder of the toolpath including line width, layer height, and deposition geometry remained unchanged across materials. This highlights the advantage of the custom toolpath generation approach, which enables consistent geometric control while allowing material-specific thermal parameters to be adjusted independently.

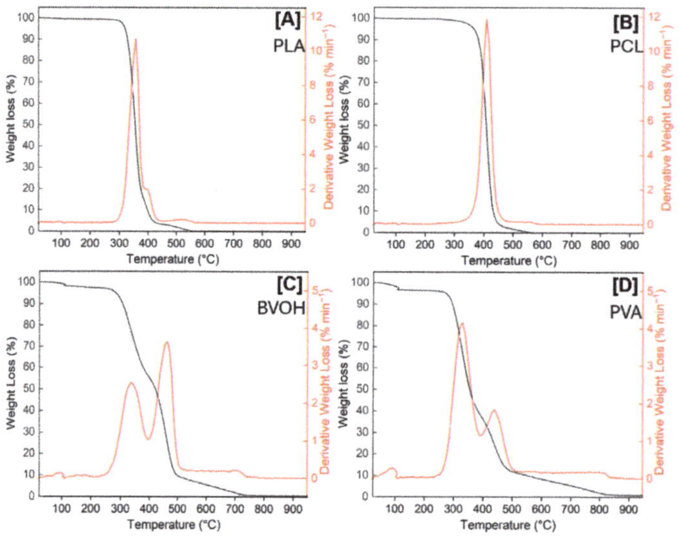

Figure 5. Comparison of thermogravimetric behaviour of PLA (**A**), PCL (**B**), BVOH (**C**), and PVA (**D**).

3.3. Structural and Chemical Analysis of Scaffolds

3.3.1. TGA

Thermogravimetric analysis (TGA) was performed to evaluate the thermal stability and degradation behaviour of PLA, PCL, BVOH, and PVA, as shown in Figure 5. The TGA curves revealed distinct mass loss profiles for each material, reflecting differences in moisture sensitivity, thermal stability, and decomposition mechanisms.

As previously mentioned, the ability to adsorb moisture (hygroscopic nature) is evident for both BVOH and PVA, which were recorded to hold 1.63 wt.% and 3.26 wt.%, respectively, whereas PLA and PCL both presented <0.3 wt.% of moisture, as reflected in the derivative weight loss plot (red) for all four polymers, with either a weight loss at ≈100 °C (Figure 5C,D) or negligible difference (Figure 5A,B). This early-stage mass is consistent with the printing challenges observed, such as produced surface roughness and filament instability.

PLA exhibited a single dominant degradation step at 360 °C, indicating good thermal stability. PCL showed a similarly well-defined degradation profile, with onset of thermal decomposition occurring at higher temperatures than PLA (412 °C). In contrast, both BVOH and PVA displayed an initial weight loss starting at lower temperatures (268 °C and 266 °C, respectively). This leads to both materials displaying two separate mass losses at

 https://doi.org/10.3390/jmmp10010039

increasing temperature due to depolymerisation; for BVOH these reached a maximum at 340 °C and 465 °C, with the latter being the major mass loss component. For PVA the two mass losses were at 331 °C and 443 °C; here the major thermal event was seen at the lower temperature, clearly demonstrating the poor thermal stability of PVA. Interestingly, PLA was the only polymer that generated an inorganic ash residue (0.8 wt.%), suggesting that the other polymers investigated were organic by nature.

Overall, the TGA results confirm that PLA and PCL possess superior thermal stability compared to BVOH and PVA, supporting their suitability as primary structural materials for 3D-printed tissue engineering scaffolds. Conversely, the lower thermal stability and moisture sensitivity of BVOH and PVA suggest they are more appropriate to use as sacrificial materials rather than load-bearing scaffold components.

3.3.2. DSC Analysis

Differential scanning calorimetry was carried out across all four polymers, the results of which are shown in Figure 6A–D; in each case the glass transition temperature (T_g), melting temperature (T_m) and material destruction temperatures (T_d) can be elucidated, the latter in close relationship with the temperatures denoted via thermogravimetric analysis. These measurements are used to infer structural alterations as temperature increases, all of which are endothermic by nature and are represented by positive Delta H values. For the cases of PLA, BVOH and PVA, there are clearly defined T_g and T_m, the former onsets being 59.93 °C, 50.78 °C and 30.06 °C, respectively. The T_m onset values were found to be 144.75 °C, 151.15 °C and 151.64 °C, respectively. For the case of PCL, it was previously reported that the T_g is ~−60 °C, whereas the T_m has been found to be ~56 °C, which is identical to the value we show in Figure 6B [39]. With this data in mind, it confirms that the operational temperatures used for both the bed and nozzle in Table 4 were appropriate, where the bed temperatures align with the T_g onset values and the nozzle temperatures are above the T_m values. The printing nozzle temperature of PCL was approximately three times its Tm, likely due to the softness of the filament. Although PCL melts at approximately 56 °C, the viscosity of the molten polymer may have remained too high for extrusion until a significantly higher temperature was reached [40,41].

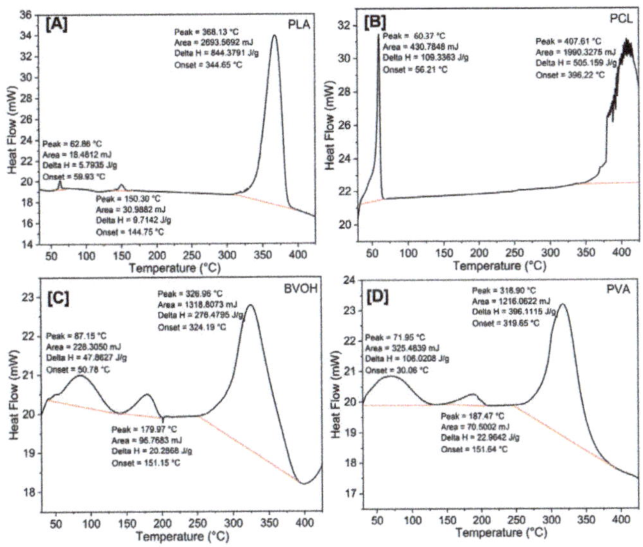

Figure 6. Comparison of DSC behaviour of PLA (**A**), PCL (**B**), BVOH (**C**), and PVA (**D**).

https://doi.org/10.3390/jmmp10010039

3.3.3. PXRD Analysis

Figure 7 compares the powder X-ray diffraction (PXRD) diffractograms of PLA, PCL, BVOH, and PVA. PCL was the only polymer that clearly showed well-defined, crystalline properties with peaks at 21.57° (110) and 24.01° (220). The other polymers are generally amorphous by nature with possible semi-crystalline properties developing at ~41°. PLA did exhibit low-intensity crystalline features at 27.65°, 36.32°, 41.36°, 44.29°, 54.54°, 56.83°, 62.97°, 64.31°, 69.20° and 69.98°, which all match TiO_2, specifically the rutile polymorph. It is believed that these were added to the PLA filament to act as a colouring agent, as shown in Figure 4A where white PLA samples are shown. The low intensity denotes a minor metal loading, which is consistent with the low ash found in the thermogravimetric analysis (<1%).

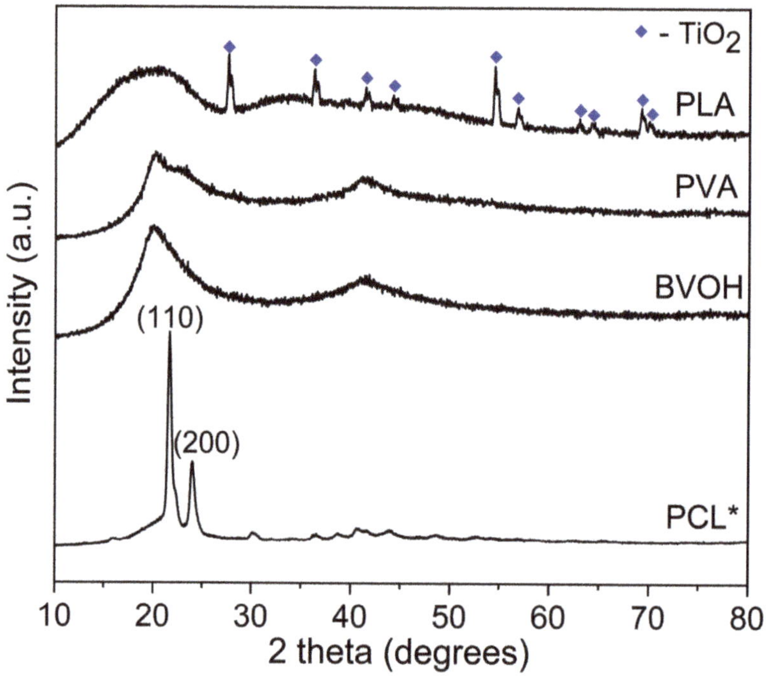

Figure 7. Comparison of PXRD behaviourPLA, PCL, BVOH, and PVA. * Due to the crystallinity of PCL, the intensity of this data was reduced by 5×.

3.4. *Tensile Strength and Porosity Analysis*

The tensile test results for all scaffold designs are presented in Figure 8, with stress (MPa) plotted against strain (Figure 8A), along with the corresponding mechanical properties, including ultimate tensile force (UTF) (Figure 8B) and Young's modulus (Figure 8C). Figure 8D illustrates scaffold clamping during tensile testing, showing improper clamping with slippage or misalignment (left) and proper clamping with secure alignment after testing (right).

As all scaffolds were fabricated from PLA, their intrinsic material properties would be identical if tested using standardised specimens (ISO 527-1:2019 [42]). This study aimed to evaluate the influence of scaffold architecture and porosity on structural performance, and therefore tensile testing was conducted using cross-sectional areas corresponding to the original scaffold geometries to determine architecture-dependent Young's modulus and ultimate tensile force (UTF). All designs were uniformly scaled by a factor of two

https://doi.org/10.3390/jmmp10010039

in the x- and y-directions while maintaining identical print parameters. The number of layers and fibres was kept constant, with the increase in size achieved by enlarging pore dimensions. Consequently, the cross-sectional area perpendicular to the applied load remained unchanged. According to Griffith's size effect [43], the reported mechanical values may therefore be conservative.

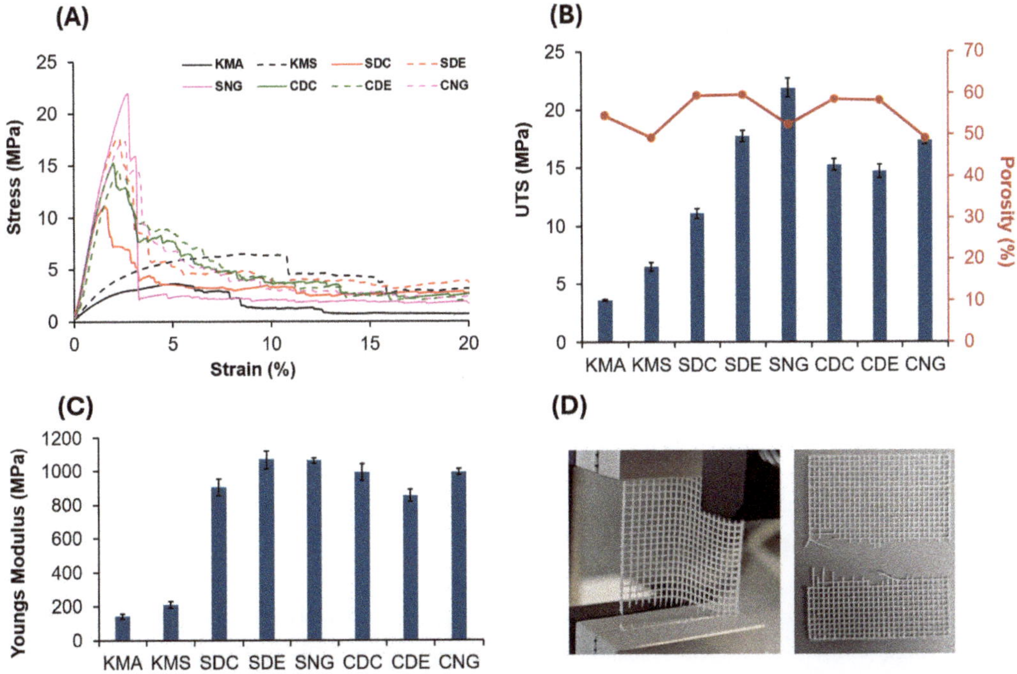

Figure 8. (**A**) Tensile test results for all scaffold designs: solid black line, knee meniscus angled (KMA); dashed black line, knee meniscus stacked (KMS); solid red line, square dense centre (SDC); dashed red line, square dense edge (SDE); solid pink line, square no gradient (SNG); solid green line, circle dense centre (CDC); dashed green line, circle dense edge (CDE); and dashed pink line, circle no gradient (CNG). (**B**) Influence of the various designs on ultimate tensile strength (UTS) and porosity. (**C**) Young's modulus. (**D**) Images illustrating scaffold clamping during tensile testing: the left image shows an improperly clamped scaffold during testing, exhibiting slippage or misalignment, while the right image shows a properly clamped scaffold after testing, demonstrating correct alignment and secure fixation within the test rig for reliable measurements.

Non-gradient designs exhibited the highest UTF values due to their smaller average pore sizes and greater number of load-bearing fibres. For example, the square no-gradient (SNG) design withstood a maximum force of 110.34 ± 0.46 N, compared to 90.53 ± 0.18 N for the circle no-gradient (CNG) design. However, these designs also exhibited lower volumetric porosities, which may limit their applicability where permeability and tissue integration are required [44,45]. Figure 8B summarises the volumetric porosity of all designs. The square dense edge (SDE) design exhibited the highest volumetric porosity ($59.63 \pm 0.79\%$), while the circle no-gradient (CNG) design showed the lowest ($49.12 \pm 1.22\%$). This difference is primarily due to fibre distribution rather than the presence of a gradient. In the SDE design, material is concentrated at the periphery, resulting in fewer fibres and larger pore volumes in the central region. In contrast, the CNG design maintains a uniform fibre distribution with consistently small pores, leading to lower overall porosity.

 https://doi.org/10.3390/jmmp10010039

Although the gradient designs did not exhibit the highest mechanical performance among the tested samples, their Young's modulus and ultimate tensile force (UTF) remained within ranges reported as suitable for tissue scaffold applications in the literature [2,46,47]. The gradient designs achieved the highest volumetric porosities, which may be advantageous in applications where biological performance is prioritised over load-bearing capacity [48,49]. High porosity facilitates nutrient diffusion, oxygen transport, and cell migration, all of which are critical for tissue ingrowth [50–52]. In addition, graded porosity can promote region-specific biological responses, supporting both surface-level healing and deeper tissue regeneration [2,38]. Gradient architecture enables spatial customisation of pore size in regions of interest while reducing material usage in areas subjected to lower mechanical demands [47,53]. This design flexibility is particularly relevant for structurally heterogeneous tissues, such as the knee meniscus. By increasing pore size locally, scaffolds may enhance wound exudate absorption and maintain a favourable healing environment, while denser regions provide mechanical support and structural guidance [47,54,55]. Among the knee meniscus designs, the knee meniscus stacked (KMS) scaffold exhibited higher mechanical performance than the knee meniscus angled (KMA) design (81.13 ± 0.251 N vs. 46.06 ± 0.114 N), which can be attributed to improved fibre alignment under tensile loading. However, this increase in strength was accompanied by reduced porosity, illustrating the trade-off between mechanical reinforcement and permeability. Conversely, the higher porosity of the KMA design may be beneficial in applications where enhanced fluid transport and cell infiltration are required. These results demonstrate that scaffold architecture can be used to tune the balance between mechanical performance and porosity, enabling design optimisation tailored to specific tissue engineering applications.

3.5. Degradation Analysis

Figure 9A shows the interpolated degradation profiles plotted as a function of time over a 60 min period. The degradation results confirm that only BVOH and PVA underwent complete dissolution in phosphate-buffered saline (PBS) under the tested conditions. BVOH exhibited the fastest degradation, with the entire scaffold dissolving within approximately 25 min. This rapid dissolution was accompanied by a visible change in the PBS solution from clear to a white, translucent appearance (Figure 9C). In contrast, the PVA scaffold dissolved more slowly, requiring approximately 60 min to fully dissolve; however, the PBS solution remained visually clear following dissolution.

The degradation also shows that PLA and PCL did not dissolve during the test period. Instead, both materials exhibited a modest initial increase in mass due to PBS absorption, resulting in slight scaffold swelling. This mass increase stabilised within approximately 20 min, with PLA and PCL absorbing approximately 6–8% of their initial mass. A two-tailed Student's t-test comparing the mass change behaviour of PLA and PCL yielded a p-value of 0.14, indicating no statistically significant difference between the two materials and confirming that they behaved similarly under the tested conditions.

BVOH and PVA also exhibited rapid initial PBS absorption prior to degradation. BVOH reached approximately 110% of its initial mass, while PVA absorbed a substantially greater volume of PBS, reaching approximately 165% of its original mass. This absorption occurred rapidly, within the first 5 min of immersion. Following this initial uptake, both materials reached a saturation point beyond which further absorption was limited. At this stage, degradation became dominant, leading to a gradual reduction in mass. As illustrated in Figure 9A, both the absorption and degradation phases for BVOH and PVA exhibited near-linear behaviour, characterised by an initial linear increase in mass, followed by a

https://doi.org/10.3390/jmmp10010039

brief plateau where absorption and degradation rates briefly achieved equilibrium, and finally a linear decrease corresponding to scaffold dissolution.

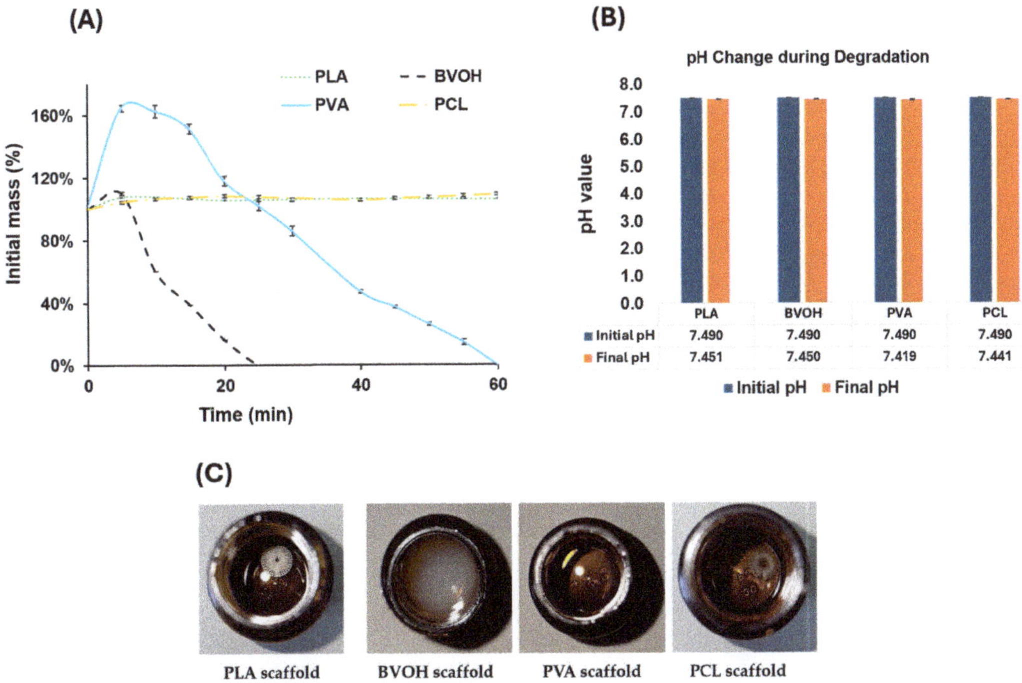

Figure 9. (**A**) Interpolated degradation rate of all materials plotted over a 60 min duration. (**B**) pH values of all four materials before and after 48 h degradation. (**C**) Photographs of PBS solutions after 48 h containing PLA, BVOH, PVA, and PCL scaffolds.

The pH was measured before and after 48 h at 37 °C, with no significant differences observed among all material types, as shown in Figure 9B. Measurement intervals were adapted to the degradation behaviour of each material; all scaffolds were initially immersed in separate PBS solutions and assessed after 30 min. As the BVOH scaffold fully dissolved within this period, subsequent measurements were taken at 5 min intervals to accurately capture the dissolution process. For PVA, which degraded more slowly, a combination of 5 min and 10 min intervals was used, with shorter intervals applied during the early and final stages of dissolution. In contrast, PLA and PCL did not dissolve, and measurements were initially taken at 5 min intervals, followed by longer intervals (10 min, 30 min, and finally 60 min) once mass stabilisation was observed. Figure 9C further illustrates the clarity of the PBS solutions following each material's degradation test, with images captured 48 h after the start of the experiment to ensure complete dissolution of any soluble degradation products.

4. Conclusions

Eight scaffold architectures:KMA, KMS, SDC, SDE, SNG, CDC, CDE, and CNGwere designed with controlled porosity gradients in the x-, y-, and z-directions and fabricated using material extrusion additive manufacturing. Tensile testing of PLA scaffolds was performed to evaluate the influence of scaffold architecture on mechanical performance, while volumetric porosity was calculated to assess mass, permeability, and suitability for tissue engineering.

https://doi.org/10.3390/jmmp10010039

All designs exhibited mechanical properties adequate for biomedical applications. However, the dense-edge designs, SDE and CDE, demonstrated the most favourable balance between mechanical strength, porosity, and mass. The SDE design achieved the highest overall performance, withstanding a maximum force of 73.4 N before fracture while maintaining a porosity of 59.63%. The CDE design showed comparable performance, with an ultimate tensile force of 60.8 N and a porosity of 58.27%. These results highlight the effectiveness of graded porosity in optimising scaffold performance. The dense-edge architectures enabled larger central pore sizes while maintaining increased material density at the periphery, allowing controlled fluid transport and improved fixation using bioadhesives. This design strategy supports mechanical stability while accommodating the biological requirements of the healing process. An in vitro degradation study using the KMA design printed in PLA, PVA, BVOH, and PCL revealed that PLA and PCL exhibited comparable degradation stability ($p > 0.05$). While both are suitable for structural scaffolds, PLA offered advantages in printability and cost. Among the water-soluble materials, PVA outperformed BVOH by dissolving more slowly, absorbing more fluid, and avoiding solution discolouration.

Overall, this study demonstrates that custom toolpath-based additive manufacturing provides a powerful platform for precisely tailoring scaffold architecture, mechanical behaviour, porosity distribution, and degradation characteristics. These findings establish a strong foundation for the development of patient-specific meniscus scaffolds and highlight the potential of architecturally graded designs to advance personalised tissue engineering and regenerative medicine.

Author Contributions: Conceptualization, A.P. and B.Z.; methodology, A.P., A.M., M.J.T., F.E.L., H.G., J.H., and B.Z.; validation, A.P., A.M., M.J.T., F.E.L., H.G., J.H., and B.Z.; formal analysis A.P., A.M., M.J.T., F.E.L., H.G., J.H., and B.Z.; investigation, A.P., A.M., M.J.T., F.E.L., H.G., J.H., and B.Z.; resources, A.P., A.M., M.J.T., F.E.L., H.G., J.H., and B.Z.; data curation, A.P., A.M., M.J.T., F.E.L., H.G., J.H., and B.Z.; writing—original draft preparation, A.P., A.M., M.J.T., F.E.L., and B.Z.; writing—review and editing, A.P., A.M., M.J.T., F.E.L., H.G., J.H., and B.Z.; visualization, A.P., A.M., M.J.T., F.E.L., H.G., J.H., and B.Z.; supervision, A.M., M.J.T., H.G., and B.Z.; project administration, A.P., A.M., and B.Z.; funding acquisition, A.M., M.J.T., H.G., and B.Z. All authors have read and agreed to the published version of the manuscript.

Funding: This research was funded by The Royal Society Research Grant (RG\R1\241133) and the EPSRC Centres for Doctoral Training Grant (EP/S023763/1).

Data Availability Statement: The original contributions presented in this study are included in the article. Further inquiries can be directed to the corresponding author.

Acknowledgments: B.Z. would like to thank The Royal Society Research Grant (RG\R1\241133) for supporting the research work and collaborations. M.J.T. and F.E.L. thank the EPSRC Centres for Doctoral Training in Offshore Wind Energy (EP/S023763/1) for F.E.L.'s studentship.

Conflicts of Interest: The authors declare no conflicts of interest.

References

1. Ni, Y.; Gao, Y.; Yao, J. Introduction to musculoskeletal system. In *Biomechanical Modelling and Simulation on Musculoskeletal System*; Springer: Berlin/Heidelberg, Germany, 2022; pp. 1–34.
2. Zhang, B.; Huang, J.; Narayan, R.J. Gradient scaffolds for osteochondral tissue engineering and regeneration. *J. Mater. Chem. B* **2020**, *8*, 8149–8170. [CrossRef]
3. Paggi, C.A.; Teixeira, L.M.; Le Gac, S.; Karperien, M. Joint-on-chip platforms: Entering a new era of in vitro models for arthritis. *Nat. Rev. Rheumatol.* **2022**, *18*, 217–231. [CrossRef]
4. Nochehdehi, A.R.; Nemavhola, F.; Thomas, S. Structure, function, and biomechanics of meniscus cartilage. In *Cartilage Tissue and Knee Joint Biomechanics*; Elsevier: Amsterdam, The Netherlands, 2024; pp. 61–73.
5. McNulty, A.L.; Guilak, F. Mechanobiology of the meniscus. *J. Biomech.* **2015**, *48*, 1469–1478. [CrossRef] [PubMed]

 https://doi.org/10.3390/jmmp10010039

6. Masouros, S.; McDermott, I.; Amis, A.; Bull, A. Biomechanics of the meniscus-meniscal ligament construct of the knee. *Knee Surg. Sports Traumatol. Arthrosc.* **2008**, *16*, 1121–1132. [CrossRef]
7. Logerstedt, D.S.; Ebert, J.R.; MacLeod, T.D.; Heiderscheit, B.C.; Gabbett, T.J.; Eckenrode, B.J. Effects of and response to mechanical loading on the knee. *Sports Med.* **2022**, *52*, 201–235. [CrossRef]
8. Orozco Grajales, G.A. *Mechanobiological Modeling of Articular Cartilage: Predicting Post-Traumatic Knee Osteoarthritis*; Itä-Suomen Yliopisto: Kuopio, Finland, 2020.
9. Zhang, B.; Chung, S.H.; Barker, S.; Craig, D.; Narayan, R.J.; Huang, J. Direct ink writing of polycaprolactone/polyethylene oxide based 3D constructs. *Prog. Nat. Sci. Mater. Int.* **2021**, *31*, 180–191. [CrossRef]
10. Yan, W.; Dai, W.; Cheng, J.; Fan, Y.; Wu, T.; Zhao, F.; Zhang, J.; Hu, X.; Ao, Y. Advances in the mechanisms affecting meniscal avascular zone repair and therapies. *Front. Cell Dev. Biol.* **2021**, *9*, 758217. [CrossRef]
11. Longo, U.G.; Campi, S.; Romeo, G.; Spiezia, F.; Maffulli, N.; Denaro, V. Biological strategies to enhance healing of the avascular area of the meniscus. *Stem Cells Int.* **2012**, *2012*, 528359. [CrossRef] [PubMed]
12. Kuczyński, N.; Boś, J.; Białoskórska, K.; Aleksandrowicz, Z.; Turoń, B.; Zabrzyńska, M.; Bonowicz, K.; Gagat, M. The Meniscus: Basic Science and Therapeutic Approaches. *J. Clin. Med.* **2025**, *14*, 2020. [CrossRef]
13. Murphy, C.A.; Garg, A.K.; Silva-Correia, J.; Reis, R.L.; Oliveira, J.M.; Collins, M.N. The meniscus in normal and osteoarthritic tissues: Facing the structure property challenges and current treatment trends. *Annu. Rev. Biomed. Eng.* **2019**, *21*, 495–521. [CrossRef]
14. Uchida, D.T.; Bruschi, M.L. 3D printing as a technological strategy for the personalized treatment of wound healing. *AAPS PharmSciTech* **2023**, *24*, 41. [CrossRef]
15. Zhao, J.; Ning, C.; Jiang, R.; Zhang, R.; Li, J.; Xie, Y.; Li, J.; Qiao, K.; Bahatibieke, A.; Lu, Q. Host–Guest Functionalized Hierarchical Porous Polyurethane Scaffold for Mechano-Biological Coupling in Meniscus Reconstruction. *Adv. Funct. Mater.* **2025**, e23756. [CrossRef]
16. Allum, J.; Moetazedian, A.; Gleadall, A.; Mitchell, N.; Marinopoulos, T.; McAdam, I.; Li, S.; Silberschmidt, V.V. Extra-wide deposition in extrusion additive manufacturing: A new convention for improved interlayer mechanical performance. *Addit. Manuf.* **2023**, *61*, 103334. [CrossRef]
17. Moetazedian, A.; Budisuharto, A.S.; Silberschmidt, V.V.; Gleadall, A. CONVEX (CONtinuously Varied EXtrusion): A new scale of design for additive manufacturing. *Addit. Manuf.* **2021**, *37*, 101576. [CrossRef]
18. Park, H.S.; Park, H.J.; Lee, J.; Kim, P.; Lee, J.S.; Lee, Y.J.; Seo, Y.B.; Kim, D.Y.; Ajiteru, O.; Lee, O.J. A 4-axis technique for three-dimensional printing of an artificial trachea. *Tissue Eng. Regen. Med.* **2018**, *15*, 415–425. [CrossRef]
19. Zhang, B.; Cristescu, R.; Chrisey, D.B.; Narayan, R.J. Solvent-based extrusion 3D printing for the fabrication of tissue engineering scaffolds. *Int. J. Bioprint.* **2020**, *6*, 211. [CrossRef]
20. Poonawala, K.; Zhang, B. Influence of 3D printing process parameters and design on mechanical properties of tissue scaffolds. *MATEC Web Conf.* **2025**, *413*, 08003. [CrossRef]
21. Winarso, R.; Anggoro, P.; Ismail, R.; Jamari, J.; Bayuseno, A. Application of fused deposition modeling (FDM) on bone scaffold manufacturing process: A review. *Heliyon* **2022**, *8*, e11701. [CrossRef]
22. Azad, M.A.; Olawuni, D.; Kimbell, G.; Badruddoza, A.Z.M.; Hossain, M.S.; Sultana, T. Polymers for extrusion-based 3D printing of pharmaceuticals: A holistic materials–process perspective. *Pharmaceutics* **2020**, *12*, 124. [CrossRef]
23. Domínguez-Robles, J.; Martin, N.K.; Fong, M.L.; Stewart, S.A.; Irwin, N.J.; Rial-Hermida, M.I.; Donnelly, R.F.; Larrañeta, E. Antioxidant PLA composites containing lignin for 3D printing applications: A potential material for healthcare applications. *Pharmaceutics* **2019**, *11*, 165. [CrossRef]
24. Tripathi, D.; Sharma, A.; Tyagi, P.; Beniwal, C.S.; Mittal, G.; Jamini, A.; Singh, H.; Tyagi, A. Fabrication of three-dimensional bioactive composite scaffolds for hemostasis and wound healing. *AAPS PharmSciTech* **2021**, *22*, 138. [CrossRef]
25. Moetazedian, A.; Candeo, A.; Liu, S.; Hughes, A.; Nasrollahi, V.; Saadat, M.; Bassi, A.; Grover, L.M.; Cox, L.R.; Poologasundarampillai, G. Versatile microfluidics for biofabrication platforms enabled by an agile and inexpensive fabrication pipeline. *Adv. Healthc. Mater.* **2023**, *12*, 2300636. [CrossRef]
26. Tran, H.Q.; Shahriar, S.S.; Yan, Z.; Xie, J. Recent advances in functional wound dressings. *Adv. Wound Care* **2023**, *12*, 399–427. [CrossRef]
27. Badekila, A.K.; Kini, S.; Jaiswal, A.K. Fabrication techniques of biomimetic scaffolds in three-dimensional cell culture: A review. *J. Cell. Physiol.* **2021**, *236*, 741–762. [CrossRef]
28. Dynamism. Eco-PLA Technical Data Sheet. Available online: https://asset.dynamism.com/media/catalog/product/pdf/Dynamism_PLA_TDS.pdf (accessed on 11 January 2026).
29. RS Components Ltd. Verbatim 1.75 mm White PLA 3D Printer Filament. Available online: https://uk.rs-online.com/web/p/3d-printing-materials/8226459 (accessed on 10 January 2026).
30. BigRep GmbH. BVOH 3D Printing Support Material. Available online: https://bigrep.com/filaments/bvoh (accessed on 10 January 2026).

https://doi.org/10.3390/jmmp10010039

31. Daemon3DPrint. Kexcelled THE K6™ BVOH Technical Data Sheet. Available online: https://www.daemon3dprint.com/media/sparsh/product_attachment/THE_K6_BVOH_TDS.pdf (accessed on 9 January 2026).
32. Verbatim. Verbatim 1.75 mm Transparent BVOH 3D Printer Filament. Available online: https://uk.rs-online.com/web/p/3d-printing-materials/2020419 (accessed on 10 January 2026).
33. Dynamism. Technical Data Sheet—PVA. Available online: https://asset.dynamism.com/media/catalog/product/pdf/Dynamism_PVA_TDS.pdf (accessed on 10 January 2026).
34. 3D4Makers. Facilan™ PCL 100 Filament. Available online: https://www.3d4makers.com/products/facilan-pcl-100-filament?srsltid=AfmBOorVOmOn5hHBSLjRPp-9saRiYylpWr2f_FXgrVsJDu6c0h8oWBSc (accessed on 10 January 2026).
35. Gleadall, A. FullControl GCode Designer: Open-source software for unconstrained design in additive manufacturing. *Addit. Manuf.* **2021**, *46*, 102109. [CrossRef]
36. Moetazedian, A.; Gleadall, A.; Han, X.; Ekinci, A.; Mele, E.; Silberschmidt, V.V. Mechanical performance of 3D printed polylactide during degradation. *Addit. Manuf.* **2021**, *38*, 101764. [CrossRef]
37. Ottenio, M.; Tran, D.; Annaidh, A.N.; Gilchrist, M.D.; Bruyère, K. Strain rate and anisotropy effects on the tensile failure characteristics of human skin. *J. Mech. Behav. Biomed. Mater.* **2015**, *41*, 241–250. [CrossRef]
38. Seijo-Rabina, A.; Paramés-Estevez, S.; Concheiro, A.; Pérez-Muñuzuri, A.; Alvarez-Lorenzo, C. Effect of wound dressing porosity and exudate viscosity on the exudate absorption: In vitro and in silico tests with 3D printed hydrogels. *Int. J. Pharm. X* **2024**, *8*, 100288. [CrossRef]
39. Baptista, C.; Azagury, A.; Shin, H.; Baker, C.M.; Ly, E.; Lee, R.; Mathiowitz, E. The effect of temperature and pressure on polycaprolactone morphology. *Polymer* **2020**, *191*, 122227. [CrossRef]
40. Paetzold, R.; Coulter, F.B.; Singh, G.; Kelly, D.J.; O'Cearbhaill, E.D. Fused filament fabrication of polycaprolactone bioscaffolds: Influence of fabrication parameters and thermal environment on geometric fidelity and mechanical properties. *Bioprinting* **2022**, *27*, e00206. [CrossRef]
41. Zhang, B.; Gleadall, A.; Belton, P.; Mcdonagh, T.; Bibb, R.; Qi, S. New insights into the effects of porosity, pore length, pore shape and pore alignment on drug release from extrusionbased additive manufactured pharmaceuticals. *Addit. Manuf.* **2021**, *46*, 102196. [CrossRef]
42. *EN ISO 527-1:1996*; Plastics—Determination of Tensile Properties: General Principles. ISO: Geneva, Switzerland, 2019.
43. Bažant, Z.P. Size effect on structural strength: A review. *Arch. Appl. Mech.* **1999**, *69*, 703–725. [CrossRef]
44. Loh, Q.L.; Choong, C. Three-dimensional scaffolds for tissue engineering applications: Role of porosity and pore size. *Tissue Eng. Part B Rev.* **2013**, *19*, 485–502. [CrossRef]
45. Hollister, S.J. Porous scaffold design for tissue engineering. *Nat. Mater.* **2005**, *4*, 518–524. [CrossRef]
46. Wahlsten, A.; Stracuzzi, A.; Lüchtefeld, I.; Restivo, G.; Lindenblatt, N.; Giampietro, C.; Ehret, A.E.; Mazza, E. Multiscale mechanical analysis of the elastic modulus of skin. *Acta Biomater.* **2023**, *170*, 155–168. [CrossRef]
47. Zhang, B.; Guo, L.; Chen, H.; Ventikos, Y.; Narayan, R.J.; Huang, J. Finite element evaluations of the mechanical properties of polycaprolactone/hydroxyapatite scaffolds by direct ink writing: Effects of pore geometry. *J. Mech. Behav. Biomed. Mater.* **2020**, *104*, 103665. [CrossRef]
48. Breish, F.; Hamm, C.; Andresen, S. Nature's load-bearing design principles and their application in engineering: A review. *Biomimetics* **2024**, *9*, 545. [CrossRef]
49. Jin, Y.; Li, J.; Fan, H.; Du, J.; He, Y. Biomechanics and Mechanobiology of Additively Manufactured Porous Load-Bearing Bone Implants. *Small* **2025**, *21*, 2409955. [CrossRef]
50. Rademakers, T.; Horvath, J.M.; van Blitterswijk, C.A.; LaPointe, V.L. Oxygen and nutrient delivery in tissue engineering: Approaches to graft vascularization. *J. Tissue Eng. Regen. Med.* **2019**, *13*, 1815–1829. [CrossRef]
51. Zhang, B. Computational fluid dynamics analysis of the fluid environment of 3D printed tissue scaffolds within a perfusion bioreactor: The effect of pore shape. *Procedia Struct. Integr.* **2023**, *49*, 3–9. [CrossRef]
52. Zhan, Z.; Wang, Y.; Xie, H.; Yang, M.; Ruan, M.; Liu, X.; Liu, J.; Liu, Z.; Wen, F.; Hong, X. Hierarchically Porous Microgels with Interior Spiral Canals for High-Efficiency Delivery of Stem Cells in Wound Healing. *Small* **2025**, *21*, 2405648. [CrossRef]
53. Ye, F.; Yang, H.; Hong, C.; Wu, X.; Dai, H. Design and characterization of 3D printed gradient scaffolds with spatial distribution of pore sizes. *Ceram. Int.* **2025**, *51*, 12185–12196. [CrossRef]
54. Zahid, A.A.; Chakraborty, A.; Shamiya, Y.; Ravi, S.P.; Paul, A. Leveraging the advancements in functional biomaterials and scaffold fabrication technologies for chronic wound healing applications. *Mater. Horiz.* **2022**, *9*, 1850–1865. [CrossRef]
55. Mukasheva, F.; Adilova, L.; Dyussenbinov, A.; Yernaimanova, B.; Abilev, M.; Akilbekova, D. Optimizing scaffold pore size for tissue engineering: Insights across various tissue types. *Front. Bioeng. Biotechnol.* **2024**, *12*, 1444986. [CrossRef]

https://doi.org/10.3390/jmmp10010039

Journal of
Manufacturing and Materials Processing

MDPI

Article

Development and Mechanical Testing of Synthetic 3D-Printed Models of Healthy and Metastatic Vertebrae

Daniela Bruno [1], Riccardo Forni [2], Marco Palanca [1,*], Luca Cristofolini [1] and Paolo Gargiulo [2,3]

1 Department of Industrial Engineering, Alma Mater Studiorum—University of Bologna, 40131 Bologna, Italy; daniela.bruno8@unibo.it (D.B.); luca.cristofolini@unibo.it (L.C.)
2 Institute of Biomedical and Neural Engineering, Reykjavik University, 102 Reykjavik, Iceland; riccardo21@ru.is (R.F.); paolo@ru.is (P.G.)
3 Department of Science, Landspitalin, University Hospital of Iceland, 105 Reykjavik, Iceland
* Correspondence: marco.palanca@unibo.it

Abstract: Experimental characterisation of ex vivo specimens is limited by specimen availability and high costs, whereas 3D printing provides a cost-effective alternative for producing multiple replicas. This study aimed to develop a methodology for evaluating the individual and combined effects of material composition and geometry on the biomechanical performance of 3D-printed vertebrae. CT scans of healthy human vertebrae and with lytic metastases were segmented to fabricate synthetic models through Digital Anatomy Printing. Three types of 3D-printed models were produced: Healthy vertebrae, Metastatic vertebrae, and Healed vertebrae (metastatic geometry filled with healthy material). All models were tested under axial compression to measure the strength, stiffness, and strain. Repeatability across replicas was assessed as well as comparison of mechanical properties among the different vertebral types. Results showed excellent repeatability, with coefficients of variation below 5% for strength and stiffness-related parameters. The Metastatic models exhibited significant reductions in strength compared to Healthy ones, while stiffness remained similar, consistent with ex vivo data trends. Healed models highlighted the role of material composition in driving mechanical behaviour, independently of geometry. This work provides the first quantitative assessment of 3D-printed vertebrae with metastatic lesions, supporting their future potential as standardised alternatives to cadaveric testing.

Keywords: 3D printing; vertebrae; metastasis; experimental test; digital anatomy printing

1. Introduction

The spine is among the most frequent sites of bone metastases (87%) [1]. These lesions are typically found in the thoracic (60–80%) and lumbar spine (15–30%) [2,3], and are predominantly located in the vertebral bodies rather than in the posterior elements [4]. Lytic metastasis is associated with the highest risk of vertebral fracture [4–8], as it affects the microarchitecture of the bone by locally reducing the bone mineral density [1,9–12].

The effects of the metastases on the biomechanics of the vertebrae have been investigated using both experimental [5–7,13] and computational studies [4,14,15]. Ex vivo experimental testing allows for direct measurement of apparent properties (e.g., stiffness, strength) and local properties (e.g., surface or volumetric strains). However, it faces challenges such as limited availability of human specimens, high costs, intra-specimen variability, and a trade-off between time-consuming procedures and tissue deterioration. Moreover, since destructive tests can only be performed once, the quantification of the mechanical properties depends on the specific applied scenario (i.e., loading rate, loading direction, etc.). As a result, these failure-based measures prevent comprehensive characterisation of the mechanical response and may hide some effects [13].

FE models allow for multiple predictions on the same specimen. However, current modelling challenges in accurately predicting the complex mechanical behaviour of metastatic vertebrae restrict their clinical application [14].

Given these challenges, the use of 3D-printed (3DP) synthetic models has emerged as a promising alternative. Additive manufacturing, or 3D printing, allows for the fabrication of customised models that reflect individual anatomical variability, using patient-specific medical images [16–18]. This technology allows for the production of large numbers of identical replicas at relatively low cost [19,20], thereby facilitating controlled and repeatable biomechanical testing under a wide spectrum of loading conditions [21,22]. Furthermore, the introduction of multi-material 3D printing allows the fabrication of objects with a wide range of heterogeneous material properties and functionalities [23], decreasing the production time without extra cost for the fabrications of complex morphologies [24]. Recently, various multi-material additive manufacturing technologies for polymers have been developed and implemented in different applications, enabling the clinical adoption of patient-specific anatomical models and guides, particularly for surgical planning and dentistry or maxillofacial procedures, where multi-material contrasts enhance visual and tactile realism and support procedural rehearsal [23,25].

Complex structures such as the cortical-like shells with trabecular-like interiors, including pathological features like the metastases, can be produced incorporating tailored materials to enhance both anatomical and mechanical realism [26–28]. Despite this capacity, the 3D printing is adopted in the biomedical field mainly for surgical simulation and anatomical reconstruction [29,30]. Its potential in biomechanical testing remains only partially explored. There is still a lack of systematic evaluation regarding the mechanical characterisation of these printed models in biomechanical testing contexts.

This study aims to develop a robust methodology for evaluating the individual and combined effects of material composition and geometry on the biomechanical performance of 3DP vertebrae. To achieve this aim, the investigation is organised across two objectives:

1. Evaluation of the repeatability of mechanical parameters across replicas of the same vertebra.
2. Assessment of the influence of material composition and geometry on the mechanical behaviour of the synthetic models emulating healthy, pathological, and treated vertebrae.

2. Materials and Methods

2.1. Clinical Data

Four spines from donors with an history of vertebral metastases were obtained through an ethically approved donation program (Anatomic Gifts Registry, Hanover, MD, USA). Five spine segments (Table 1) consisting in one vertebra with lytic metastases and one adjacent healthy vertebra were prepared. The spine segments were imaged with a Computed Tomography (CT) (AquilonOne, Toshiba, Japan) following an optimised bone imaging protocol [6,13] (voltage: 120 kVp, current: 200 mA, slice thickness: 1 mm, in-plane resolution: around 0.45 mm, matrix size: 512 × 512).

Table 1. Clinical data of the donors, level of the healthy, and metastatic vertebrae.

ID	Donor	Primary Tumour	Sex	Age (yr)	BMI (kg/m^2)	Healthy Vertebra	Metastatic Vertebra
#1	A	Nasopharyngeal	M	72	16	T7	T6
#2	B	Breast	F	82	22	T7	T6
#3	B	Breast	F	82	22	L3	L4
#4	C	Adenocarcinoma	F	62	57	T6	T5
#5	D	Breast	F	55	17	T7	T8

2.2. Segmentation

Each CT scan in 16-bit was imported into Mimics software (version 26.0; Materialise, Leuven, Belgium) to enable construction of a 3D virtual model. The vertebral bodies and the metastatic lesions were segmented by the same expert operator (DB) to ensure consistency across all samples. A standardised protocol was followed (Figure 1): firstly, the bone tissue (cortical and trabecular) was segmented using a semi-automated Hounsfield Unit (HU) threshold-based approach with a range of 200 to 1780 HU [31,32]. Then, a Region of Interest (ROI) was defined by isolating the segmented bone tissue of the two central adjacent vertebrae, the healthy and the metastatic ones. After that, a 3D region-growing algorithm was applied to each vertebra included in the ROI, followed by an automatic hole-filling tool to refine the bone tissue mask.

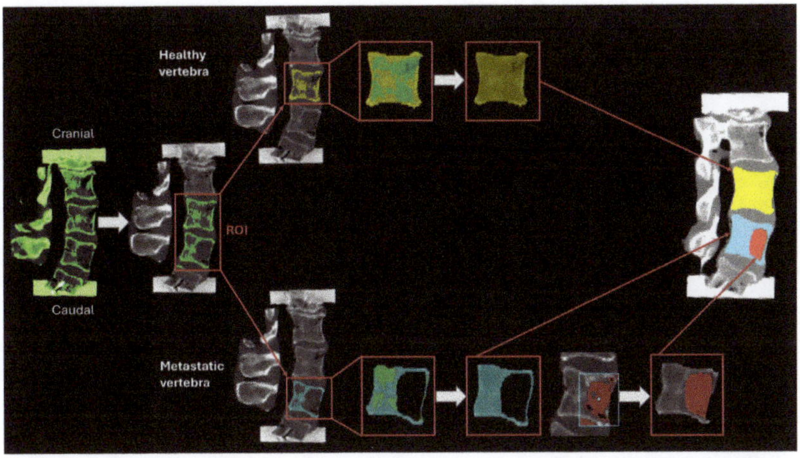

Figure 1. CT-based segmentation protocol. From the left, a global bone mask (green) is generated using a threshold of [200–1780] HU, and the ROI (healthy and metastatic vertebra) is defined. In the upper row, region growing is applied to extract the healthy vertebra (yellow). In the lower row, region growing extracts the metastatic vertebra (blue), holes are filled, and the metastasis is additionally segmented (red). On the right, the final segmentation displays the three components together.

The metastatic lesion was segmented using the same semi-automated protocol (HU: $-100 \div 200$). A Boolean operation was applied to avoid overlaps between the bone and lesion tissue, followed by a visual inspection of each 2D axial slice to verify the accuracy of this separation. Subsequently, the segmented masks underwent mesh smoothing (5 iterations; smoothing factor: 0.4) to eliminate surface artefacts and mesh wrapping (smallest detail size: 0.5 mm; gap-closing distance: 0.25 mm) to establish a uniform cortical layer of 1–2 mm thick and minimise the risk of stress concentrations.

After the segmentation, the cross-sectional area and volume of each vertebra were calculated. The cross-sectional area was measured using the CT scans as the minimum cross-section area of the vertebra. In the case of metastatic vertebrae, the metastasis was included.

The volume was calculated for the healthy vertebrae, the metastatic vertebrae (excluding the volume of the metastasis), and the metastases from the CT scans. The metastasis volume was then expressed as a percentage of the total volume of the vertebral body. Lastly, the 3D virtual models were exported in Standard Tessellation Language (STL) format for 3D printing.

2.3. Material Assignment

Material assignment within each vertebra model was managed using GrabCAD Digital Anatomy Creator (DAC) 1.73 software (Stratasys, Eden Prairie, MN, USA), which enables voxel-level control over material distribution to mimic the heterogeneous architecture of biological tissues [33].

Two material blends were selected to emulate the mechanical behaviour and structural composition of the vertebrae and metastases: MatBone (preset Dense Vertebra, DAC, Stratasys), designed to emulate the bone tissue, and MatMet (preset Highly Contractible, DAC, Stratasys), intended to emulate the lytic lesions (Figure 2). MatBone was recommended by the manufacturer for bone-mimicking applications. The preset provides a dense outer shell and a more porous inner structure, analogous to the cortical and trabecular regions, respectively. Specifically, MatBone had a layered structure consisting of:

A. An outer layer (0.35 mm) made entirely of BoneMatrix (BM), an elastoplastic polymer with a Young's modulus of around 250 MPa [28]. This layer provides shape memory and elastic recovery properties.
B. A transitional layer (1.65 mm) made from a blend of 90% BM and 10% of VeroClear (VC) (Young's modulus $\approx$ 500 MPa) [28].
C. An inner porous layer composed of VC combined with a wax-support material (SUP706) (Young's modulus $\approx$ 200 MPa), designed to emulate the architecture of trabecular bone.

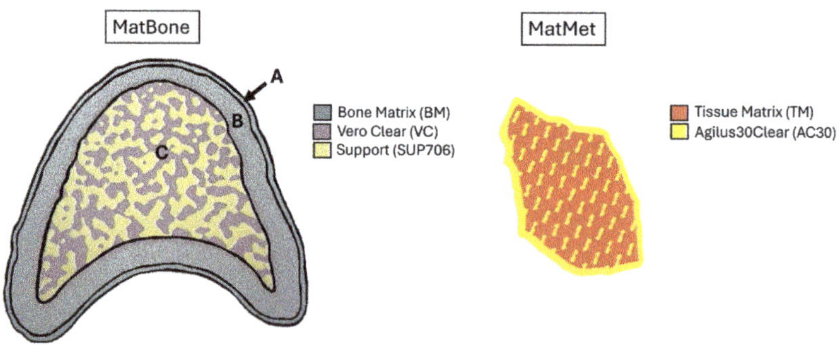

Figure 2. On the left: material blend MatBone, with layers A (outer layer), B (transitional layer), C (inner layer), in which blue represents BM, purple VC and yellow SUP706. On the right: material blend MatMet, in which red represents TM and yellow AC30.

MatMet was chosen for printing the metastatic tissue due to its lower-stiffness and elastoplastic behaviour (Young's Modulus $\approx$ 0.3 MPa) [28,34] typical of the lytic metastases [35,36]. MatMet featured a gyroid infill pattern, a type of triply periodic minimal surface (TPMS) lattice structure, composed of a 1:1 blend of TissueMatrix (TM), a soft-gel and highly deformable elastomer (Shore 00 = 30), and Agilus30Clear (AC30) [28].

2.4. 3DP Models

All the 3DP vertebra models were fabricated using a Stratasys J850 Digital Anatomy Printer (Stratasys, Eden Prairie, MN, USA), a multi-material jetting 3D printing system based on PolyJet technology [26]. Each STL model was digitally sliced into thin layers (0.014 mm) in which the printer selectively deposits microscopic droplets of photopolymer material, either single resin or blend. Following deposition, each layer was immediately cured by ultraviolet (UV) light exposure, and the build platform was incrementally lowered to enable sequential layer-by-layer construction [37].

Three types of 3DP vertebral models were fabricated (Figure 3):

- Healthy vertebrae, which had the geometry of the healthy vertebrae and were printed entirely with MatBone.
- Metastatic vertebrae, which had the geometry of the metastatic vertebrae and were printed using MatBone for the bone structure and MatMet for the metastatic lesions.
- Healed vertebrae, which had the same geometry as the metastatic models, but they were made exclusively of MatBone, thereby replacing the material used for the metastatic tissue with the material used for the healthy tissue.

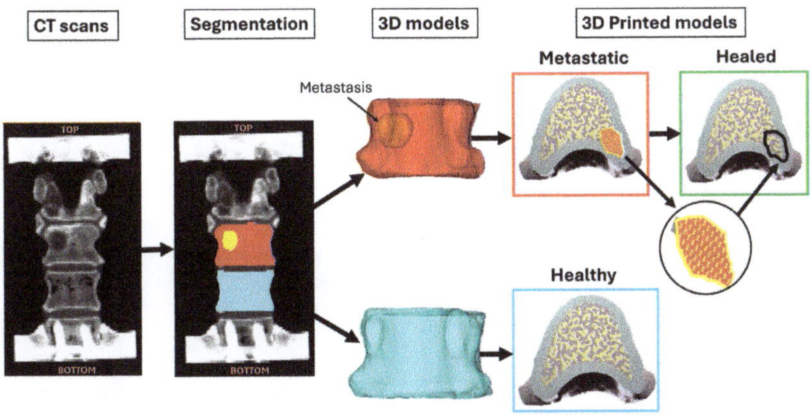

Figure 3. Workflow for printing 3D vertebral models. From left to right: coronal CT slice; segmented masks showing the metastatic vertebra (red), metastasis (yellow), and healthy vertebra (blue); 3D virtual model of the metastatic vertebra (top) with metastasis inside, and healthy vertebrae (bottom); corresponding 3DP models, with internal structure highlighted in cross section for the three model types (Healthy, Metastatic, Healed).

For each vertebra model (Healthy, Metastatic, and Healed), three replicas were printed resulting in a total of 45 specimens across five distinct spinal segments. In order to avoid local stress concentrations arising from the contact between the irregular shape of the vertebral endplates and the loading plates, and to ensure a uniform load-bearing surface during mechanical testing, all 3DP vertebral models were placed between VC cylindrical pots, which had dimensions of 10 mm in height and 50 mm in diameter (Figure 4). The pots were designed using commercial Computer-Aided Design (CAD) software (PTC Creo 7.0, PTC Inc., Boston, MA, USA) and featured internal cavities that were precisely matched to the vertebra endplates geometry to ensure a precise fit. To avoid any interfacial gaps or misalignments that could impair load transmission, both the vertebrae and the cylinders were printed as one integrated structure simultaneously. By using this method, there was perfect continuity between the contact surfaces, which minimised the risk of stress concentration or interface separation during testing.

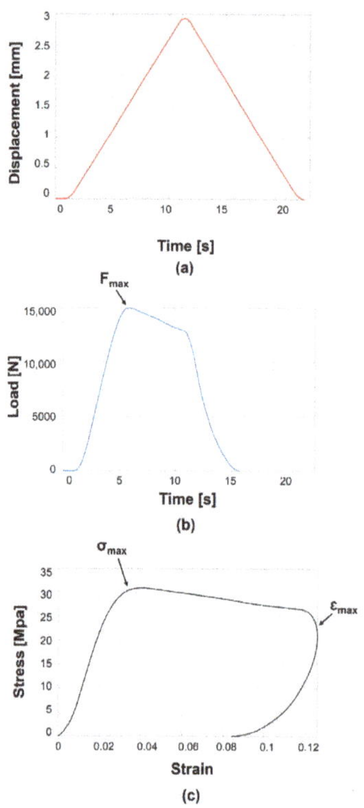

Figure 4. Mechanical test loading protocol: (**a**) imposed displacement-time curve, (**b**) load-time curve for a typical specimen, (**c**) stress-strain curve for a typical specimen, (**d**) mechanical setup.

2.5. Mechanical Testing Protocol and Metrics

Each 3DP vertebra was tested under axial compression. A uniaxial servo-hydraulic testing machine (Instron 8500 controller, Instron, Buckinghamshire, UK), equipped with a 100 kN load cell and a custom loading setup comprising two parallel flat plates, was used (Figure 4). The mechanical testing protocol consisted of a displacement-controlled trapezoidal load ramp: involving a 3 mm displacement applied at a rate of 0.3 mm/s, followed by a holding phase of 1 s, and an unloading ramp at 0.3 mm/s (Figure 4). This load-hold-unload profile was chosen to assess elastic and viscoelastic response characteristics of the printed models. For each 3DP model, force and displacement signals were acquired using a multichannel data acquisition system (PXIe, National Instruments, Austin, TX, USA). Signals were acquired at a frequency of 100 Hz and filtered using a 50-point moving average to minimise high-frequency noise. To isolate the mechanical response of the vertebral body, the displacement was corrected by subtracting the initial offset and the contribution of the supporting cylindrical pots. For each pot was estimated the height reduction associated to the compression, using the following formula:

$$\Delta = \frac{F \times h}{E \times A}$$

where Δ is the displacement of the pot, F is the applied force, h is the height of the pot, E is the Young's Modulus of the VC (Young's Modulus $\approx$ 1500 MPa), and A is the cross-section of the cylindrical pot.

The height reduction of the two cylinders was subtracted from the measured test displacement. Finally, the corrected displacement was used to calculate the mechanical properties of the vertebrae.

Based on force-displacement data, four apparent properties were computed: maximum force (F_{max}), maximum stress (σ_{max}), maximum strain (ε_{max}), and Young's modulus (E). Stress (σ) was calculated by dividing the applied force by the minimum cross-sectional area of the respective vertebra. Strain (ε) was calculated as the displacement divided by the height of the vertebral body. Young's modulus (E) was calculated as the slope of the linear regression between 20% and 80% of the maximum stress on the loading portion of the stress–strain curve. The range between 20% and 80% of the maximum stress represents the linear elastic region of the stress–strain curve, which excludes the initial adjustments of the testing machine [38].

2.6. Statistical Analysis

To evaluate measurements repeatability among the three replicas of the same vertebra, the coefficient of variation (CV%) was calculated, for each parameter, as follows:

$$CV(\%) = 100 \times \frac{\text{Standard deviation}}{\text{Mean}}$$

A CV below 5% was considered indicative of good repeatability [39].

Subsequently, data normality and homogeneity of variance were assessed using the Shapiro-Wilk test and Levene's test, respectively. Based on these results, non-parametric tests were used for F_{max}, σ_{max}, and ε_{max}, while parametric tests were applied for E. Comparisons between the different types of vertebrae were performed, accordingly:

- To evaluate the material effect, Metastatic vs. Healed types were compared. The two groups were treated as paired, as both models were obtained from the same vertebrae but printed them in different ways. Therefore, the Wilcoxon signed-rank test was applied to F_{max}, σ_{max}, and ε_{max}, while paired *t*-test was applied to E.
- To evaluate the geometry effect, Healthy vs. Healed types were compared. The two groups were treated as independent since they are based on different vertebrae, therefore, the Mann-Whitney U test was used to F_{max}, σ_{max}, and ε_{max}, while an independent *t*-test was applied to E.
- To evaluate the combined effect of material and geometry, Healthy vs. Metastatic types were compared. They were considered independent, hence the Mann-Whitney U test was used to F_{max}, σ_{max}, and ε_{max}, while an independent *t*-test was applied to E.

All statistical analyses were performed in Matlab (MATLAB R2022a, The MathWorks Inc., Natick, MA, USA) with a level of significance of 0.05.

3. Results

3.1. Geometrical Parameters

The volume of the Healthy vertebrae ranged from 11,645 mm^3 to 25,810 mm^3 (Table 2). In comparison, Metastatic vertebrae showed smaller volumes, ranging from 10,059 mm^3 to 23,185 mm^3, except for specimen #5 in which the Metastatic vertebra had a slightly larger volume than the healthy one (13,980 mm^3 vs. 13,970 mm^3). Healed vertebrae exhibited volumes ranging from 10,585 mm^3 to 26,991 mm^3, exceeding those of Healthy vertebrae in specimens #3 and #5.

Table 2. Geometrical parameters of each vertebra in each segment: minimum cross-sectional area, volume, and percentage of metastatic lesion in the vertebral body (VB).

ID	Typologies	Minimum Cross-Sectional Area (mm^2)	Volume of the Vertebra (mm^3)	Volume of the Metastasis (mm^3) (VB%)
#1	Healthy	532	11,645	-
	Metastatic	486	10,059	526 (4.9)
	Healed	486	10,585	-
#2	Healthy	570	12,363	-
	Metastatic	493	10,481	737 (6.6)
	Healed	493	11,218	-
#3	Healthy	848	25,810	-
	Metastatic	828	23,185	3806 (14.1)
	Healed	828	26,991	-
#4	Healthy	606	14,323	-
	Metastatic	524	10,280	696 (6.3)
	Healed	524	10,976	-
#5	Healthy	598	13,970	-
	Metastatic	623	13,980	912 (6.1)
	Healed	623	14,892	-

The size of lytic metastases ranged from 526 mm^3 to 3806 mm^3, corresponding to 4.9% to 14.1% of the total volume of the vertebral body. All lesions were focal and localised in the vertebral body, with no clear prevalence of a particular position observed across specimens.

The minimum cross-sectional area of the vertebrae followed a similar trend: healthy vertebrae ranged from 532 mm^2 to 848 mm^2, while Metastatic vertebrae, and consequently also Healed vertebrae ranged from 486 mm^2 to 828 mm^2. Similarly, in specimen #5, the Metastatic vertebra showed a larger area (623 mm^2) than the corresponding Healthy one (598 mm^2).

3.2. *Repeatability of Mechanical Parameters on Vertebrae Replicas*

All four biomechanical parameters (F_{max}, σ_{max}, ε_{max}, and E) were evaluated for all specimens across the three vertebral types (Healthy, Metastatic, Healed) (Figure 5 and Table 3).

Regarding F_{max}, the Healthy type showed a median value of approximately 17.4 kN, with a distribution ranging 14.8 kN to 19.7 kN, the Metastatic specimens a median of 15 kN with a range 13.4 to 17.2 kN, the Healed specimens a median of 19.5 kN with a range 15.5 kN to 24.3 kN.

For σ_{max}, the Healthy specimens displayed a median value of approximately 28.0 MPa with a range 23.2 to 31.4 MPa, the Metastatic specimens a median of 26.5 MPa with a range 20.1 MPa to 31.2 MPa, the Healed type had a median value of 33.5 MPa with a range 29.3 MPa to 37.9 MPa).

ε_{max} values were relatively consistent across the three typologies, with only minor variations in both median and range. The median strain value observed for the Healthy type was 0.12, for the Metastatic type 0.13, and for the Healed type 0.12.

E for the Healthy type showed a median of 1259 MPa and ranged from 626 to 1259 MPa, for Metastatic type the median was 1294 MPa and ranged 561 to 1294 MPa and for Healed type the median was 1400 MPa and ranged 595 to 1400 MPa.

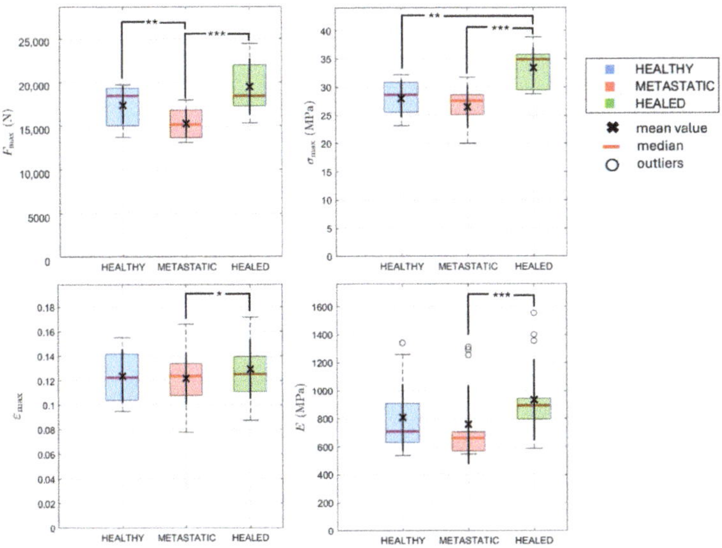

Figure 5. Distribution of maximum force (F_{max}), maximum stress (σ_{max}), maximum strain (ε_{max}) and Young's modulus (E) across the three vertebral typologies: Healthy (blue), Metastatic (red), and Healed (green). The data are represented as box plots, where the red line indicates the median, the black cross denotes the mean, and circles identify outliers, defined as values beyond 1.5 times the interquartile range (IQR). Whiskers extend to the most extreme data points within this range. Each type includes n = 15 specimens. Statistical significance levels are indicated as follows: * $p < 0.05$, ** $p < 0.01$, *** $p < 0.001$.

Table 3. Median values and range for each biomechanical parameter, calculated across replicas for each spine segment and typology (Healthy, Metastatic, Healed).

	ID	F_{max} (kN)	σ_{max} (MPa)	ε_{max}	E (MPa)
Healthy	#1	15.2 (13.7–16.2)	28.7 (25.8–30.5)	0.14 (0.14–0.14)	1259 (1006–1499)
	#2	14.8 (14.6–15.0)	26.1 (25.6–26.3)	0.12 (0.10–0.14)	655 (643–664)
	#3	19.7 (19.6–19.7)	23.2 (23.2–23.3)	0.10 (0.09–0.11)	626 (626–627)
	#4	18.7 (18.2–19.5)	30.8 (30.1–32.2)	0.15 (0.14–0.16)	708 (537–742)
	#5	18.8 (18.5–18.9)	31.4 (30.9–31.6)	0.11 (0.10–0.12)	906 (871–908)
Metastatic	#1	15.2 (15.0–15.4)	31.2 (30.8–31.7)	0.12 (0.12–0.13)	1294 (1258–1323)
	#2	13.7 (13.7–13.8)	27.8 (27.7–28.0)	0.12 (0.10–0.13)	699 (695–706)
	#3	16.7 (16.6–17.0)	20.1 (20.0–20.5)	0.09 (0.08–0.11)	561 (546–568)
	#4	13.4 (13.1–13.7)	25.6 (25.0–26.1)	0.14 (0.14–0.17)	572 (549–584)
	#5	17.2 (16.9–18.0)	27.6 (27.2–28.9)	0.12 (0.11–0.14)	661 (658–695)
Healed	#1	18.5 (17.9–18.9)	37.9 (36.8–38.9)	0.12 (0.12–0.13)	1400 (1328–1499)
	#2	17.6 (17.2–17.6)	35.7 (34.9–35.8)	0.13 (0.09–0.14)	942 (933–941)
	#3	24.3 (23.8–24.4)	29.3 (28.8–29.5)	0.10 (0.10–0.11)	803 (791–807)
	#4	15.5 (15.3–15.9)	29.5 (29.2–30.4)	0.17 (0.16–0.17)	595 (589–620)
	#5	21.8 (21.5–22.0)	35.0 (34.6–35.4)	0.13 (0.12–0.14)	893 (873–900)

Across the five healthy vertebrae, F_{max} and σ_{max} exhibited CV between replicas consistently below 5% (Table 4), except for specimen #1, which showed a CV of 8.4%. ε_{max} showed slightly more variability, with CVs ranging from 1.2% to 19.0%. The highest value was observed in specimen #2. E values ranged from 0.1% to 16.5%, with specimens #1 and #4 exceeding the 5% threshold. Among the metastatic vertebrae, F_{max} and σ_{max} exhibited low variation, with CV values between 0.6% and 3.2%. While ε_{max} showed greater variability, ranging from 0.6% to 15.4%. In contrast, E exhibited excellent repeatability, with all values remaining below 3%. Healed vertebrae followed a similar trend to the metastatic vertebrae: F_{max} and σ_{max} presented CV values below 2.7%, while ε_{max} exhibited increased variability. In particular, specimen #2 reached a CV of 23.3%, the highest across all typologies. E values were below 3% in all specimens, except for specimen #1, which had a CV of 7.2%.

Table 4. Coefficient of variation (CV%) for each biomechanical parameter, calculated across replicas for each spine segment and typology (Healthy, Metastatic, Healed). Bold values indicate values above 5%.

	ID	F_{max}	σ_{max}	ε_{max}	E
Healthy	#1	**8.4%**	**8.4%**	1.2%	**14.6%**
	#2	1.4%	1.4%	**19.0%**	1.5%
	#3	0.2%	0.2%	**5.7%**	0.1%
	#4	3.5%	3.5%	**7.0%**	**16.5%**
	#5	1.1%	1.1%	**6.0%**	2.5%
Metastatic	#1	1.5%	1.5%	0.6%	2.3%
	#2	0.6%	0.6%	**10.9%**	1.0%
	#3	1.3%	1.3%	**15.4%**	2.0%
	#4	2.1%	2.1%	**9.7%**	2.9%
	#5	3.2%	3.2%	**9.4%**	2.8%
Healed	#1	2.7%	2.7%	0.9%	**7.2%**
	#2	1.3%	1.3%	**23.3%**	0.6%
	#3	1.3%	1.3%	3.0%	1.0%
	#4	2.0%	2.0%	3.5%	2.7%
	#5	1.1%	1.1%	**6.0%**	1.3%

Overall, all three types demonstrated excellent repeatability in strength-related metrics (F_{max} and σ_{max}) and stiffness. Strain measures, instead, showed lower consistency.

3.3. *Influence of Material Composition and Geometry on the Mechanical Behaviour*

Pairwise statistical analyses showed significant differences among the 3DP vertebrae types for most parameters (Figure 5).

The Metastatic type showed a 17% reduction of the F_{max} (Mann-Whitney U test, $p < 0.01$) with respect to Healthy type and a 19% reduction (Wilcoxon test, $p < 0.001$) with respect to the Healed type. No significant difference was observed between Healthy and Healed types (Mann-Whitney U test, $p = 0.16$).

The Metastatic specimens showed a 23% reduction of the σ_{max} (Wilcoxon test, $p < 0.001$) compared to the Healed type. While the Healthy type showed a 20% reduction of the σ_{max} (Mann-Whitney U test, $p < 0.01$) with respect to the Healed type. No significant difference was found between Metastatic and Healthy types (Mann-Whitney U test, $p = 0.40$).

The Metastatic type showed a 8% reduction of the ε_{max} (Wilcoxon test, $p < 0.05$) compared to the Healed type. No statistically significant differences were detected between Healthy and Metastatic types (Mann-Whitney U test, $p = 0.80$) or between Healthy and Healed types (Mann-Whitney U test, $p = 0.67$).

Finally, the Metastatic type showed a reduction of 34% of the E (paired t test, $p < 0.001$) with respect to the Healed type. Conversely, no statistical differences were found between Healthy and Healed types (independent t-test, $p = 0.21$) or between Healthy and Metastatic types (independent t-test, $p = 0.60$).

4. Discussion

This study aimed to create a reliable method for assessing the biomechanical performance of 3DP vertebrae models. The objectives were to: (1) determine the repeatability of mechanical parameters among vertebra replicas, and (2) examine how material composition and geometry influence the mechanical behaviour of various vertebrae.

A total of forty-five vertebrae, emulating Healthy, Metastatic, and Healed conditions, were reconstructed from CT scans, 3D-printed, and mechanically tested.

Our findings demonstrate that 3DP models comply with the mechanical trends observed in human vertebrae, supporting their use in future biomechanical investigations, either as a complement to or an alternative to ex vivo testing. Indeed, the repeatability analysis confirmed excellent consistency in strength-related metrics (F_{max} and σ_{max}) and stiffness across replicas, with CV values below the 5% threshold (Table 4) [39]. This low variability confirmed good manufacturing repeatability, ensuring identical geometry and controlled materials. This enables multiple destructive tests on replicas of the same vertebra geometry, which is not otherwise feasible with ex vivo specimens [13,40]. On the other hand, higher variability was observed for strain measurements (ε_{max}), which can be attributed to the material viscoelastic behaviour and the lack of preconditioning before testing.

The comparative analysis among the vertebral types revealed clear trends: the metastatic models exhibited significant reductions in the maximum force compared to the healthy ones, while their Young's modulus remained similar (Figure 5). This trend is consistent with ex vivo studies, where lytic metastases reduce vertebral strength while having a limited effect on the global stiffness [6].

In general, the strength and the elastic properties of 3DP models were higher in absolute terms than those reported for human vertebrae in the literature. This difference could be expected due to the elastoplastic behaviour and the homogeneous nature of the used polymeric materials, which lack the viscoelastic behaviour and trabecular microstructure typical of real bone tissue. In the healthy models, the median F_{max} reached 19.5 kN (range: 15.5–24.3 kN), compared to 2.2 $\pm$ 0.9 kN reported by [5] and 6.2 $\pm$ 3.0 kN predicted by [41]. In the same way, the Metastatic models (median: 15.2 kN; range: 13.4–17.2 kN) exceeded ex vivo measurements (1.4 $\pm$ 0.7 kN [5], 1.2 $\pm$ 0.9 kN [7], 5.8 kN (range: 3.0–6.2 kN) [15]), and in silico prediction (6.1 $\pm$ 3.5 kN [41]). In this study, the σ_{max} exhibited trends similar to those of F_{max}, but with absolute values higher than those reported in the literature (Healthy vertebra: 28.0 MPa vs. 7.7 MPa; Metastatic vertebra: 26.5 MPa vs. 4.5 MPa [42]). For ε_{max}, the 3DP models exhibited medians of 0.13 in Metastatic types and 0.12 in Healthy ones, compared to 0.008 $\pm$ 0.005 for metastatic and 0.003 $\pm$ 0.009 for healthy in ex vivo human vertebrae [6]. Although the absolute ε_{max} values are substantially higher, the proportional increase in strain from Healthy to Metastatic models nonetheless parallels the trends reported in the literature. In terms of Young's modulus, the 3DP models decreased from 626–1259 MPa in the Healthy vertebrae to 561–1294 MPa in the Metastatic vertebrae, consistent with ex vivo findings: ref. [5] reported 382 $\pm$ 194 MPa (healthy) versus 250 $\pm$ 130 MPa (metastatic), and ref. [42] reported 855 $\pm$ 360 MPa versus 667 $\pm$ 121 MPa. The Healed models, which preserve the geometry of the Metastatic vertebra but replace the metastasis tissue with healthy material, showed significant increases in all biomechanical parameters compared to both Metastatic and Healthy vertebrae (Figure 5), highlighting the role of material properties in determining vertebral mechanical properties. In addition, the use of 'Healed' models provided an unique opportunity to isolate the effects of material composition and, due to its identical size, a valuable comparison to the metastatic condition [13]. This approach overcomes a common limitation in ex vivo studies where adjacent vertebrae, frequently used as controls, which often differ in bone density, trabecular architecture, and endplate morphology, thereby confounding interpretation exhibit differing mechanical properties [13]. On the other hand, maximum stress (σ_{max}) was significantly influenced by the geometry, likely due to its dependence on the minimum cross-sectional area used in its calculation. Instead, maximum force (F_{max}), was conditioned by both material properties and geometry. This pattern may be explained by the higher load-bearing capacity and more uniform stress distribution of the rigid polymer used to simulate healthy trabecular bone, compared to the softer pathological material emulating the lytic lesions.

To the best of our knowledge, quantitative assessments of the strength and stiffness of 3DP vertebrae, particularly in the presence of metastatic lesions, were missing. Previous works have only focused on the visual and tactile feedback of synthetic models [43,44].

This study has some limitations. First, the available printable materials do not fully replicate the anisotropy and viscoelasticity of trabecular and cortical bone, nor those of the metastatic lesions. Although Digital Anatomy Printing allows the reproduction of heterogeneous structures with voxel-level control [33], the mechanical properties of the resulting models still differ from those of biological tissues (Figure 5). Second, some variability may have been introduced during the segmentation phase. Despite the use of a standardised HU thresholding protocol and the fact that all segmentations were performed by an experienced operator, minor variations in CT slice thickness, threshold selection, or user-dependent operation could lead to subtle geometric deviations. Furthermore, segmentation of bone metastases is challenging due to the lack of a validated protocol and by the significant overlap of HU values between tumoral tissue and native bone marrow [15,33]. As highlighted by [45], segmentation remains a critical source of geometric inaccuracies in medical 3D-printing workflows, particularly when modelling complex morphologies such as metastatic lesions. Third, mechanical testing was limited to uniaxial compression. While this setup facilitates direct comparisons across specimens, it does not reproduce the complex multi-axial loading conditions (e.g., lateral bending or flexion) typically experienced by the vertebrae [13]. Finally, the study included a relatively small sample size (n = 5). A power analysis was performed to determine the required sample size. Assuming that vertebrae with lytic metastases exhibit approximately 50% of the strength of healthy vertebrae ($\approx$2500 N [6]), with a variability of 1000 N in healthy vertebrae, a Type I error rate of 5%, and a Type II error rate of 20%, a minimum of five specimens per group was sufficient to detect statistically significant differences. Nevertheless, a larger sample size is mandatory to assess the inter-specimen variability of the 3DP models.

Future studies should focus on tuning the properties of artificial materials on biological ones to better emulate the mechanical behaviour of biological tissue. In addition, expanding mechanical testing to include different loading conditions and incorporating spinal segments with multiple vertebrae, intervertebral discs, and ligaments. This would better replicate the complex physiological conditions.

5. Conclusions

In this work, we have quantified for the first time the mechanical properties of synthetic 3DP models of vertebrae. This approach demonstrated excellent repeatability and confirmed that the mechanical performance of the 3DP models is highly dependent on material properties. This study provides a unique framework to isolate the effects of metastatic lesions in vertebrae. The observed trends mirrored those of human vertebrae, with Metastatic models showing a reduction in strength compared to healthy ones but a similar Young's modulus. This study supports the potential of the 3DP models as a cost-effective alternative to ex vivo for biomechanical testing, as well as a promising tool for preclinical evaluation.

Author Contributions: Conceptualization, D.B., R.F., M.P., L.C. and P.G.; methodology, D.B., R.F. and M.P.; software, R.F. and P.G.; validation, R.F., M.P. and D.B.; formal analysis, D.B.; investigation, D.B., M.P. and R.F.; resources, R.F. and M.P.; data curation, D.B. and M.P.; writing—original draft preparation, D.B.; writing—review and editing, D.B., M.P., R.F. and L.C.; visualization, D.B.; supervision, M.P., L.C. and P.G.; project administration, L.C. and P.G. All authors have read and agreed to the published version of the manuscript.

Funding: The study was partially supported by the AOSpine Discovery and Innovation Awards (AOSDIA 2019_063_TUM_Palanca, 2019), Marie Skłodowska-Curie Individual Fellowship (MetaSpine, MSCA-IF-EF-ST, 832430/2018, 2018) and by the METASTRA project (EU H2022 grant ID 101080135).

Data Availability Statement: The original contributions presented in this study are included in the article. Further inquiries can be directed to the corresponding author.

Acknowledgments: The authors thank Enrico Dall'Ara for kindly sharing the clinical images.

Conflicts of Interest: The authors declare no conflicts of interest.

Abbreviations

The following abbreviations are used in this manuscript:

3D	Three-dimensional
FE	Finite Element
CT	Computed Tomography
HU	Hounsfield Unit
ROI	Region of Interest
STL	Standard Tessellation Language
UV	Ultraviolet
DAC	Digital Anatomy Creator
BM	BoneMatrix
VC	VeroClear
SUP706	Support
TPMS	Triply Periodic Minimal Surface
TM	TissueMatrix
AC30	Agilus30Clear
3DP	Three-dimensional Printed
CAD	Computer-Aided Design
CV	Coefficient of Variation
VB	Vertebral Body

References

1. Coleman, R.E.; Croucher, P.I.; Padhani, A.R.; Clézardin, P.; Chow, E.; Fallon, M.; Guise, T.; Colangeli, S.; Capanna, R.; Costa, L. Bone Metastases. *Nat. Rev. Dis. Primer* **2020**, *6*, 83. [CrossRef] [PubMed]
2. Perrin, R.G.; Laxton, A.W. Metastatic Spine Disease: Epidemiology, Pathophysiology, and Evaluation of Patients. *Neurosurg. Clin. N. Am.* **2004**, *15*, 365–373. [CrossRef]
3. Sutcliffe, P.; Connock, M.; Shyangdan, D.; Court, R.; Kandala, N.-B.; Clarke, A. A Systematic Review of Evidence on Malignant Spinal Metastases: Natural History and Technologies for Identifying Patients at High Risk of Vertebral Fracture and Spinal Cord Compression. *Health Technol. Assess.* **2013**, *17*, 42. [CrossRef]
4. Galbusera, F.; Qian, Z.; Casaroli, G.; Bassani, T.; Costa, F.; Schlager, B.; Wilke, H.-J. The Role of the Size and Location of the Tumors and of the Vertebral Anatomy in Determining the Structural Stability of the Metastatically Involved Spine: A Finite Element Study. *Transl. Oncol.* **2018**, *11*, 639–646. [CrossRef]
5. Alkalay, R.N. Effect of the Metastatic Defect on the Structural Response and Failure Process of Human Vertebrae: An Experimental Study. *Clin. Biomech.* **2015**, *30*, 121–128. [CrossRef]
6. Palanca, M.; Cavazzoni, G.; Dall'Ara, E. The Role of Bone Metastases on the Mechanical Competence of Human Vertebrae. *Bone* **2023**, *173*, 116814. [CrossRef]
7. Palanca, M.; Barbanti-Bròdano, G.; Cristofolini, L. The Size of Simulated Lytic Metastases Affects the Strain Distribution on the Anterior Surface of the Vertebra. *J. Biomech. Eng.* **2018**, *140*, 111005. [CrossRef]
8. Randall, R.L. (Ed.) *Metastatic Bone Disease: An Integrated Approach to Patient Care*; Springer International Publishing: Cham, Switzerland, 2024; ISBN 978-3-031-52000-6.
9. Whyne, C.M. Biomechanics of Metastatic Disease in the Vertebral Column. *Neurol. Res.* **2014**, *36*, 493–501. [CrossRef]
10. Macedo, F.; Ladeira, K.; Pinho, F.; Saraiva, N.; Bonito, N.; Pinto, L.; Gonçalves, F. Bone Metastases: An Overview. *Oncol. Rev.* **2017**, *11*, 321. [CrossRef] [PubMed]

11. Bailey, S.; Stadelmann, M.A.; Zysset, P.K.; Vashishth, D.; Alkalay, R.N. Influence of Metastatic Bone Lesion Type and Tumor Origin on Human Vertebral Bone Architecture, Matrix Quality, and Mechanical Properties. *J. Bone Miner. Res.* **2020**, *37*, 896–907. [CrossRef] [PubMed]
12. Cavazzoni, G.; Dall'Ara, E.; Palanca, M. Microstructure of the Human Metastatic Vertebral Body. *Front. Endocrinol.* **2025**, *15*, 1508504. [CrossRef]
13. Palanca, M.; Barbanti-Bròdano, G.; Marras, D.; Marciante, M.; Serra, M.; Gasbarrini, A.; Dall'Ara, E.; Cristofolini, L. Type, Size and Position of Metastatic Lesions Explain the Deformation of the Vertebrae under Complex Loading Conditions. *Bone* **2021**, *151*, 116028. [CrossRef] [PubMed]
14. Costa, M.C.; Campello, L.B.B.; Ryan, M.; Rochester, J.; Viceconti, M.; Dall'Ara, E. Effect of Size and Location of Simulated Lytic Lesions on the Structural Properties of Human Vertebral Bodies, a Micro-Finite Element Study. *Bone Rep.* **2020**, *12*, 100257. [CrossRef] [PubMed]
15. Stadelmann, M.A.; Schenk, D.E.; Maquer, G.; Lenherr, C.; Buck, F.M.; Bosshardt, D.D.; Hoppe, S.; Theumann, N.; Alkalay, R.N.; Zysset, P.K. Conventional Finite Element Models Estimate the Strength of Metastatic Human Vertebrae despite Alterations of the Bone's Tissue and Structure. *Bone* **2020**, *141*, 115598. [CrossRef]
16. Tejo-Otero, A.; Buj-Corral, I.; Fenollosa-Artés, F. 3D Printing in Medicine for Preoperative Surgical Planning: A Review. *Ann. Biomed. Eng.* **2020**, *48*, 536–555. [CrossRef]
17. Kabra, A.; Mehta, N.; Garg, B. 3D Printing in Spine Care: A Review of Current Applications. *J. Clin. Orthop. Trauma* **2022**, *35*, 102044. [CrossRef] [PubMed]
18. Dolino, G.; Coato, D.; Forni, R.; Boretti, G.; Ciliberti, F.K.; Gargiulo, P. Designing a Synthetic 3D-Printed Knee Cartilage: FEA Model, Micro-Structure and Mechanical Characteristics. *Appl. Sci.* **2023**, *14*, 331. [CrossRef]
19. Tack, P.; Victor, J.; Gemmel, P.; Annemans, L. 3D-Printing Techniques in a Medical Setting: A Systematic Literature Review. *Biomed. Eng. OnLine* **2016**, *15*, 115. [CrossRef]
20. Serrano, C.; Fontenay, S.; Van Den Brink, H.; Pineau, J.; Prognon, P.; Martelli, N. Evaluation of 3D Printing Costs in Surgery: A Systematic Review. *Int. J. Technol. Assess. Health Care* **2020**, *36*, 349–355. [CrossRef]
21. Metzner, F.; Neupetsch, C.; Carabello, A.; Pietsch, M.; Wendler, T.; Drossel, W.-G. Biomechanical Validation of Additively Manufactured Artificial Femoral Bones. *BMC Biomed. Eng.* **2022**, *4*, 6. [CrossRef]
22. Wahbeh, J.M.; Hookasian, E.; Lama, J.; Alam, L.; Park, S.; Sangiorgio, S.N.; Ebramzadeh, E. An Additively Manufactured Model for Preclinical Testing of Cervical Devices. *JOR SPINE* **2024**, *7*, e1285. [CrossRef] [PubMed]
23. Zheng, Y.; Zhang, W.; Baca Lopez, D.M.; Ahmad, R. Scientometric Analysis and Systematic Review of Multi-Material Additive Manufacturing of Polymers. *Polymers* **2021**, *13*, 1957. [CrossRef] [PubMed]
24. Rafiee, M. Multi Material 3D and 4D Printing: A Survey. *Adv. Sci.* **2020**, *7*, 1902307. [CrossRef]
25. Tee, Y.L.; Peng, C.; Pille, P.; Leary, M.; Tran, P. PolyJet 3D Printing of Composite Materials: Experimental and Modelling Approach. *JOM* **2020**, *72*, 1105–1117. [CrossRef]
26. Ghimire, A.; Chen, P.-Y. Mechanical Properties of Additively Manufactured Multi-Material Stiff-Soft Interfaces: Guidelines to Manufacture Complex Interface Composites with Tunable Properties. *Mater. Des.* **2024**, *238*, 112677. [CrossRef]
27. Forni, R.; Calderonel, D.; Coato, D.; Dolino, G.; Cesarelli, G.; Ricciardi, C.; Cesarelli, M.; Gargiulo, P. Towards Bio-Mimetic 3D Printable Human Anatomies. In Proceedings of the 2024 IEEE International Conference on Metrology for eXtended Reality, Artificial Intelligence and Neural Engineering (MetroXRAINE), St Albans, UK, 21 October 2024; pp. 1016–1021.
28. García-Collado, A.; Blanco, J.M.; Gupta, M.K.; Dorado-Vicente, R. Advances in Polymers Based Multi-Material Additive Manufacturing Techniques: State-of-Art Review on Properties and Applications. *Addit. Manuf.* **2022**, *50*, 102577. [CrossRef]
29. Jacobs, S.; Grunert, R.; Mohr, F.W.; Falk, V. 3D-Imaging of Cardiac Structures Using 3D Heart Models for Planning in Heart Surgery: A Preliminary Study. *Interact. Cardiovasc. Thorac. Surg.* **2008**, *7*, 6–9. [CrossRef]
30. Cui, J.; Chen, L.; Guan, X.; Ye, L.; Wang, H.; Liu, L. Surgical Planning, Three-Dimensional Model Surgery and Preshaped Implants in Treatment of Bilateral Craniomaxillofacial Post-Traumatic Deformities. *J. Oral Maxillofac. Surg.* **2014**, *72*, 1138.e1–1138.e14. [CrossRef]
31. Hangartner, T.N. Thresholding Technique for Accurate Analysis of Density and Geometry in QCT, pQCT and ÌCT Images. *J. Musculoskelet Neuronal Interact.* **2007**, *7*, 9–16.
32. Van Eijnatten, M.; Van Dijk, R.; Dobbe, J.; Streekstra, G.; Koivisto, J.; Wolff, J. CT Image Segmentation Methods for Bone Used in Medical Additive Manufacturing. *Med. Eng. Phys.* **2018**, *51*, 6–16. [CrossRef] [PubMed]
33. Heo, H.; Jin, Y.; Yang, D.; Wier, C.; Minard, A.; Dahotre, N.B.; Neogi, A. Manufacturing and Characterization of Hybrid Bulk Voxelated Biomaterials Printed by Digital Anatomy 3D Printing. *Polymers* **2020**, *13*, 123. [CrossRef] [PubMed]
34. Forni, R.; Pavan, G.; Gunnarsson, A.; Ricciardi, C.; Corsi, C.; Gargiulo, P. Advanced 3D Printing of Patient-Specific Human Heart for Improved Surgical Planning. In Proceedings of the 2023 IEEE International Conference on Metrology for eXtended Reality, Artificial Intelligence and Neural Engineering (MetroXRAINE), Milano, Italy, 25 October 2023; pp. 473–478.

35. La Barbera, L.; Villa, T.; Costa, F.; Boschetti, F.; De Robertis, M.; Anselmi, L.; Capo, G.; Pancetti, S.; Fornari, M. Poroelastic Characterization of Human Vertebral Metastases to Inform a Transdisciplinary Assessment of Spinal Tumors. *J. Clin. Med.* **2025**, *14*, 2913. [CrossRef]
36. Whyne, C.M.; Hu, S.S.; Workman, K.L.; Lotz, J.C. Biphasic Material Properties of Lytic Bone Metastases. *Ann. Biomed. Eng.* **2000**, *28*, 1154–1158. [CrossRef] [PubMed]
37. Yap, Y.L.; Yeong, W.Y. Shape Recovery Effect of 3D Printed Polymeric Honeycomb: This Paper Studies the Elastic Behaviour of Different Honeycomb Structures Produced by PolyJet Technology. *Virtual Phys. Prototyp.* **2015**, *10*, 91–99. [CrossRef]
38. Farsi, A.; Pullen, A.D.; Latham, J.P.; Bowen, J.; Carlsson, M.; Stitt, E.H.; Marigo, M. Full Deflection Profile Calculation and Young's Modulus Optimisation for Engineered High Performance Materials. *Sci. Rep.* **2017**, *7*, 46190. [CrossRef] [PubMed]
39. Machin, D.; Campbell, M.J.; Walters, S.J. *Medical Statistics*; Wiley-Blackwell: Hoboken, NJ, USA, 2007.
40. Cook, D.; Julias, M.; Nauman, E. Biological Variability in Biomechanical Engineering Research: Significance and Meta-Analysis of Current Modeling Practices. *J. Biomech.* **2014**, *47*, 1241–1250. [CrossRef]
41. Costa, M.C.; Eltes, P.; Lazary, A.; Varga, P.P.; Viceconti, M.; Dall'Ara, E. Biomechanical Assessment of Vertebrae with Lytic Metastases with Subject-Specific Finite Element Models. *J. Mech. Behav. Biomed. Mater.* **2019**, *98*, 268–290. [CrossRef]
42. Kaneko, T.S.; Bell, J.S.; Pejcic, M.R.; Tehranzadeh, J.; Keyak, J.H. Mechanical Properties, Density and Quantitative CT Scan Data of Trabecular Bone with and without Metastases. *J. Biomech.* **2004**, *37*, 523–530. [CrossRef]
43. Brouwers, L.; Teutelink, A.; Van Tilborg, F.A.J.B.; De Jongh, M.A.C.; Lansink, K.W.W.; Bemelman, M. Validation Study of 3D-Printed Anatomical Models Using 2 PLA Printers for Preoperative Planning in Trauma Surgery, a Human Cadaver Study. *Eur. J. Trauma Emerg. Surg.* **2019**, *45*, 1013–1020. [CrossRef]
44. Molinari, L.; Falcinelli, C. On the Human Vertebra Computational Modeling: A Literature Review. *Meccanica* **2022**, *57*, 599–622. [CrossRef]
45. Juergensen, L.; Rischen, R.; Hasselmann, J.; Toennemann, M.; Pollmanns, A.; Gosheger, G.; Schulze, M. Insights into Geometric Deviations of Medical 3d-Printing: A Phantom Study Utilizing Error Propagation Analysis. *3D Print. Med.* **2024**, *10*, 38. [CrossRef] [PubMed]

Journal of
Manufacturing and Materials Processing

MDPI

Article

Additively Produced Ti-6Al-4V Osteosynthesis Devices Meet the Requirements for Tensile Strength and Fatigue

Alisdair R. MacLeod [1,2], Matthew Bishop [1,2], Alberto Casonato Longo [2], Alborz Shokrani [1], Chris Rhys Bowen [1] and Harinderjit Singh Gill [1,3,*]

1 Department of Mechanical Engineering, University of Bath, Bath BA2 7AY, UK; a.macleod@bath.ac.uk (A.R.M.); mbishop@insightsurgery.com (M.B.); asc28@bath.ac.uk (A.S.); msscrb@bath.ac.uk (C.R.B.)
2 Orthoscape, 3D Metal Printing Ltd., Innovation Centre, Bath BA1 1UD, UK; alberto@orthoscape.uk
3 CBio, University of Bath, Bath BA2 7AY, UK
* Correspondence: r.gill@bath.ac.uk

Abstract: The purpose of this study was to estimate the peak stresses in a laser powder bed fusion (LPBF) additive-manufactured (AM) osteosynthesis plate during physiological loading and establish if the mechanical properties of LPBF titanium alloy were suitable for this use case. Finite element models of subject-specific osteosynthesis plates for a cohort of 28 patients were created and used to calculate the peak maximum principal stresses during physiological loading, which was estimated to be 166 MPa twelve weeks post-operatively. All specimens were LPBF additively manufactured in Ti-6Al-4V alloy. ISO compliant tests were performed for tensile and fatigue, respectively. Fatigue testing was performed for specimens that had been heat-treated only and those that had been heat-treated and polished. The Upper Yield Stress was 1012.5 ± 19.2 MPa. The fatigue limit was 227 MPa for heat-treated only specimens and increased to 286 MPa for heat-treated and polished specimens. The finite element predicted stresses were below the experimentally established limits of yield and fatigue. The tensile and fatigue properties of heat-treated LPBF Ti-6Al-4V are therefore sufficient to meet the mechanical requirements of osteosynthesis plates. Polishing is recommended to improve fatigue resistance.

Keywords: additive manufacture; titanium alloy; medical device; material properties

1. Introduction

Additive manufacturing (AM, hereafter and otherwise referred to as 3D Printing) has led to numerous innovations in orthopaedic surgery. These span from planning and surgical guides produced from polymers to implants printed with metal alloys [1–5]. AM technology has also enabled tailored regenerative solutions for controlled drug release [6]. Most AM processes are delivered through a layer-by-layer manufacturing strategy. Consequently, internal, overhanging and otherwise topologically complex geometry becomes manufacturable. In addition, the reduction of the need for complex machine programming, custom tooling and fixturing makes AM an attractive manufacturing process when production runs are short or even personalised on a part-by-part basis. As such, this family of processes has provided the opportunity for personalised custom implants, specifically where complex reconstruction is required [7,8].

Mainstream orthopaedic applications have often focused on creating implants with ingrowth surfaces for enhanced biological fixation; these have been used for acetabular (hip) [9,10] and proximal tibial (knee) [11] replacement implants. Metal AM implants have been used for revision joint replacement components [12], as the enhanced biological fixation is advantageous when dealing with limited bone stock, while primary implants are now becoming mainstream.

The metals used for additively produced implants include cobalt-chrome, tantalum and titanium alloys. The majority of contemporary AM implants are made from titanium alloys [13] and, in particular, Ti-6Al-4V ELI (ELI—extra low interstitials). This alloy is termed ASTM F136 [14] grade 23 (specifically for surgical implant applications) and conforms to the composition described in ASTM F3001–14 [15] and ISO 5832-3 [16]. Extensive studies have been performed on the biocompatibility of AM titanium alloy components [17–21]. These studies include both selective laser melting (SLM) and electron beam melting (EBM) manufacturing processes. For both AM processes, there is a large amount of data to support the low cytotoxicity of Ti-6Al-4V [22].

As well as joint replacement implants, AM has also been used for fracture fixation or osteosynthesis devices. Fractures are a common reason for hospital admission worldwide [23]; thus, osteosynthesis implants have considerably greater application than those designed for joint replacement. The use of AM technology was initially for the creation of planning guides for complex or non-standard cases [2,8,24,25]. Given the anatomic variability of bone morphology, AM has considerable potential for personalised osteosynthesis plates.

Osteosynthesis plates are also used for joint-preserving osteotomy surgery, particularly for knee osteoarthritis in the relatively young. Since a relatively young age at the time of surgery has been shown to be a major risk factor for early revision of a joint replacement [26,27], there is considerable clinical interest in delaying joint replacement. High tibial osteotomy (HTO) is a well-established procedure for treating knee osteoarthritis [28].

There is little published information on the performance of AM Ti-6Al-4V for orthopaedic osteosynthesis devices. Whereas acetabular and tibial components are principally loaded in compression [29,30], osteosynthesis plates experience bending and fail by fatigue-related mechanisms [31], typically in the presence of delayed or non-union [32,33]. The loading of HTO plates is complex and dependent upon the physiological activity being considered and the healing stage of the osteotomy. AM HTO plates provide an opportunity to provide personalised solutions for patients, which requires establishing that the AM material has appropriate properties for this use case. The key objectives of this study were to use finite element modelling to calculate plate loading during physiological activity and measure the mechanical properties of SLM-manufactured medical-grade Ti-6Al-4V ELI specimens (conforming to ASTM B348 grade 23) by a series of tensile and fatigue tests. Manufacturing defects in AM-produced parts can lead to a reduction in fatigue performance [34], as fatigue is an important failure mechanism for this type of device [35,36]. The material test specimens were inspected using SEM and nanoindention.

2. Materials and Methods

2.1. Finite Element (FE) Modelling

Computed Tomography (CT) scan data of the knee and proximal tibia from 28 subjects with knee osteoarthritis (OA) were used to create a set of proximal tibial geometries to form a cohort for an in silico trial comparing personalised with generic HTO procedures using the difference in maximum Von Mises stress in the HTO plates as the primary outcome, the full details are provided by MacLeod et al. [37]. The mean age of the subjects was 68 years (range 50 to 87 years), 54% were female, the mean body weight was 90.1 kg (range 68.8 to 121.4 kg), mean height was 1.69 m (range 1.47 to 1.90 m). For the current study, we focused on the principal stresses determined by the finite element analysis, as tensile stresses play a key role in crack propagation leading to fatigue failure under cyclic loading [38]. To facilitate interpretation of the FE results Figure A1 in the Appendix provides an overview of the whole model, exemplar physiological loading and the boundary condition, the variation in mesh density used for the various parts of the model and the locations where contact between the plates and screws was modelled.

2.1.1. Geometries and Meshing

For each subject, the CT data were segmented to obtain the proximal tibial geometry (ScanIP version M-2017.06, Synopsys, Sunnyvale, CA, USA). An open wedge high tibial osteotomy was created such that the modified mechanical axis passed through a point 62.5% [39] of the distance from the medial to lateral aspects of the tibial plateau (ANSYS SpaceClaim R18.2, ANSYS Inc., Canonsburg, PA, USA). A subject-specific plate geometry designed to match the surface profile of the subject's tibia (TOKA, Orthoscape, 3D Metal Printing Ltd, Bath, UK) was created using specialised surgical planning software (Renishaw plc, Wotton-Under-Edge, UK). Screws were modelled as cylinders with a shaft outer diameter of 5.0 mm. The plate and screws were considered to be made from Ti-6Al-4V ELI alloy.

Experimentally validated meshing parameters were used [33]. A mesh convergence study was performed [37], quadratic tetrahedral elements were used (element size 0.8 mm for plate, screws and cortical hinge, 1.4 mm for bone, 2 mm for callus, there was local refinement around plate/screw interaction to 0.3 mm).

2.1.2. Material Properties

Linear elastic isotropic material properties were used (Ti-6Al-4V: Young's modulus or E = 119 GPa, Poisson's ration or $\nu = 0.34$). For the bone, CT-based heterogeneous linear material properties were calculated using BoneMat 3.2 [40,41]. Values for E for bone ranged from 350 MPa to 23 GPa. The focus of this work was on the loads in the plate, we considered it important to represent the distribution of bone material properties rather than just considering single values for the cortical and cancellous bone as the physiological loads will be transmitted to the plate via the screw/bone interface The Young's modulus of the callus varied with healing stage [42,43], four stages were considered: Healing Stage 1 immediately post-operation with no callus formed, Healing Stage 2 (HS2) representing 2 weeks with callus E = 1.4 MPa, Healing Stage 3 (HS3) representing 6 weeks with callus E = 24 MPa and Healing Stage 4 (HS4) representing 12 weeks with callus E = 528 MPa. As clinical rehabilitation protocols limit the weight bearing immediately post-operation, Healing Stage 1 was not modelled; the finite element analysis was performed for HS2, HS3 and HS4. Poisson's ratio was taken as $\nu = 0.3$ for the bone and the callus during healing stages 2 to 4 [42,43].

2.1.3. Contact Interactions

As the interest of the study was primarily plate stress, the screw-bone interaction was modelled as tied for simplification [44]. The screw-plate interface was modelled with normal contact stiffness of 0.002 and a standard Coulomb friction model with a coefficient of friction equal to 0.8 for the tangential behaviour.

2.1.4. Loading & Boundary Conditions

Five instances in each of three physiological activities (load-steps 1 to 5: fast walking, load-steps 6 to 10: chair rise and load-steps 11 to 15: squat) were modelled, with muscle and joint loads calculated using a musculoskeletal model [45] and scaled by each subject's body weight and registered to their anatomy. The most distal part of the tibial shaft was constrained in all directions, and no other restraint was applied.

2.1.5. Solver/Solution

The FE models were solved using geometric non-linearity with an implicit solver (Ansys 18.2). The main output variables of interest in the current study were the stresses in the plate, the von Mises (σ_v), the maximum (σ_1) and minimum (σ_3) principal stresses were extracted. The peak values of these stresses over the whole cohort and all loading steps were calculated for each healing stage, HS2, HS3 and HS4.

2.2. Production of Tensile Test Specimens

The powder used to manufacture all test specimens was Ti-6Al-4V ELI (supplied by Renishaw plc, Wotton-under-Edge, UK) conforming to ASTM F136 [14] grade 23 and ASTM F3001–14 [15]. The supplier's specification gave the powder particle size as between 15 μm and 45 μm [46]. The test specimens were additively manufactured using the SLM process using an AM250 machine (Renishaw plc, Wotton-under-Edge, UK). Laser power was set to 200 W and the layer thickness to 40 μm. The following laser parameters were used: 80 μm point distance and 50 μs exposure time for borders; 60 μm point distance and 50 μs exposure time for fill hatching. The tensile testing specimens were positioned vertically on the build plate so that the potentially weaker build direction (Z) would be tested.

Previous work has raised concerns about residual stresses in AM-produced Ti-6Al-4V components [47]. To alleviate residual stress, the specimens were heat-treated in an inert argon environment, heated to 350 °C over 60 min and held for 30 min, followed by heating to 850 °C over 60 min and held for 60 min. Specimens were allowed to cool slowly.

2.3. Tensile Testing

The tensile test specimens (n = 10) were cylindrical with a reduced working section (Figure 1), and the design of the specimens complied with the proportions stated in ISO 6892-1:2019—Metallic materials—Tensile testing [48]. The tensile test specimen geometry was created in CAD (ANSYS SpaceClaim R18.2, ANSYS Inc., PA, USA) from which the stereolithography file for LPBF additive manufacturing was created. The diameter of the working section of each test specimen was measured (150mm Digital Caliper 0.01 mm with UCAS calibration, Mitutoyo (UK) Ltd., Andover, Hamps., UK) three times (measure M1, M2 & M3 pre-test) before testing, the average diameter was used to calculate the specimen specific cross-sectional area.

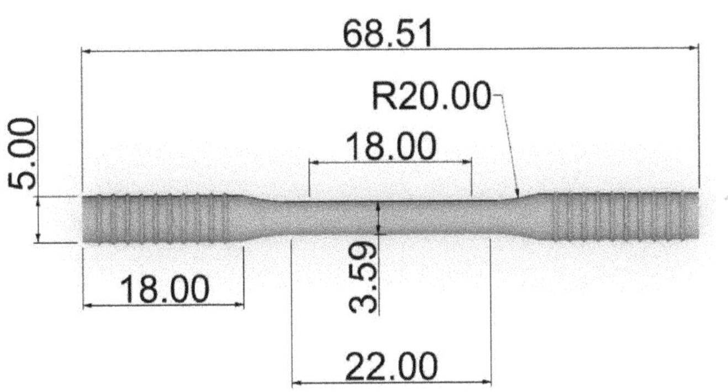

Figure 1. Specification of the tensile test samples. All dimensions in mm (not to scale).

Tensile tests to specimen failure were performed to determine the Upper Yield Strength (R_{eH}) of the additively manufactured titanium specimens according to ISO 6892-1:2019. The testing method conformed to ISO 6892-1:2019 A2. The tests were conducted using an electromechanical testing machine (INSTRON 5900 Series, fitted with a 30 kN load cell,

Instron, High Wycombe, UK). The machine applied a continuous tensile force through a predefined strain rate ($\dot{e}_{Lc}$ = 0.00025 s^{-1} corresponding to 0.005 mm/s) as per ISO 6892-1:2019 A2. Extension was measured using an extensometer (GL50, Instron, High Wycombe, UK); extension, load and the displacement of the crosshead were recorded at 10 Hz using Bluehills software (v3, Instron, High Wycombe, UK). The maximum yield stress, R_{eH}, and corresponding strain, as well as the 0.2% proof strength, Rp, and Young's Modulus, E, were determined. Note that the individual specimen cross-sectional areas were used for the stress calculations.

After the completion of each test, the specimens were again measured. The cross-section diameter was measured at the narrowest intact cross-section on the parallel length, and the final gauge length was measured using the marks from the extensometer grips.

2.4. Fatigue Test Specimens

Fatigue test specimens were also additively manufactured. These were again cylindrical with a reduced working section and made to the specification required for the fatigue testing machine used (Figure 2). The specimen geometry was again created in CAD (ANSYS SpaceClaim R18.2) from which the stereolithography file was created. Two sets of samples were produced with different finishing methods: Group 1 was heat-treated and polished (Ra = 0.32 ± 0.06 μm [49], n = 16); Group 2 were heat-treated only (Ra = 5 to 7 μm [46], n = 16). All samples were measured at the narrowest point to determine the actual diameter (repeated three times). The average (mean ± SD) diameter value for each group was 3.43 ± 0.08 mm and 3.70 ± 0.02 mm for Group 1 (polished) and Group 2, respectively, and these values were used for the calculation of bending stress.

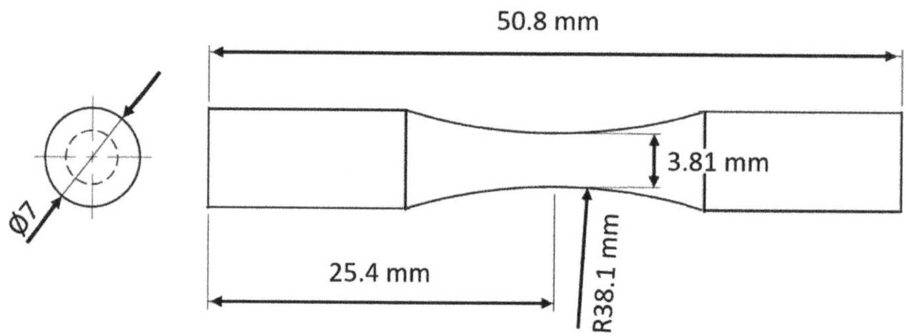

Figure 2. Specification of the fatigue test samples. All dimensions in mm (not to scale).

2.5. Fatigue Testing

The tests were conducted using a Wohler-type rotating fatigue machine (020313 RT, G.T.G. Engineering Co. Limited, Loughborough, UK). This machine applies a bending moment while simultaneously revolving the sample, causing total stress reversal in the specimen (stress ratio equal to −1) every revolution. When the sample fractured completely, the machine stopped automatically. The test machine was also used to record the time that the sample began and stopped revolving. The speed of the revolutions was measured using a tachometer (Standard AT-6, wavelength 630–670 nm, Premier Farnell Ltd., Leeds, UK). The total number of revolutions to fracture was determined by multiplying the total test time by the revolution speed.

The test machine was able to apply bending moments up to 6.8 Nm (with 68 main divisions, minimum increment was 0.1 Nm). The bending stress on the specimen, σ, was obtained from the formula:

$$\sigma = \frac{M}{Z}$$

where M is the applied bending moment and Z is the second moment of area of the specimen at the point of failure, defined by:

$$Z = \frac{\pi d}{32}$$

where d is the diameter of the specimen at the point of failure, which occurs at the smallest cross-section. For each group, the mean diameter given above was used.

The minimum number of samples recommended by ISO 12107:2012 [50] is fourteen; eight for the determination of the slope and six for evaluating the fatigue limit. The number of specimens selected for the present study was 16 for each group of specimens (Groups 1 and 2), with 9 being dedicated to determination of the sloping portion of the Stress-Number of revolutions (S-N) curve. This enabled the study to detect a 13.5% probability of failure with a confidence level of 90% ($\alpha = 0.1$) (Equation (3), Section 5.3, ISO 12107:2012 [50]). The values of bending moment and stress chosen for the sloping portion of the S-N curve are shown in Table 1. To assess repeatability, testing was repeated three times for bending moments of 2.2 Nm, 1.9 Nm and 1.6 Nm corresponding to stress levels shown in Table 1.

Table 1. Values of bending moment, resulting bending stress and number of samples selected to determine the sloping portion of the S-N curve.

Bending Moment (Nm)	Group 1 Bending Stress (MPa)	Group 2 Bending Stress (MPa)	Number of Repeats
2.2	557.3	442.0	3
1.9	481.3	381.8	3
1.6	405.3	321.5	3

The remaining seven specimens from each group were utilised for evaluating the fatigue strength with the staircase method (ISO 12107:2003), beginning with 1.1 Nm (determined from pilot tests) and adjusting in increments of 0.1 Nm; if a specimen failed, the applied stress was reduced by this increment for the next specimen, in case of non-failure (run out) the applied stress was increased by this increment for the next specimen. Fatigue tests which exceeded 1 million cycles were considered as non-failures or run outs. The fatigue strengths and standard deviations were calculated as per ISO 12107:2003. In this method, the fatigue strength is obtained by averaging the test stresses beyond the first test.

2.6. Microscopy

Two of each of the tensile and fatigue test specimens were sectioned, polished and etched as per Vander Voort [51]. These were imaged using a digital microscope (VHX-6000, Keyence Ltd., Milton Keynes, UK).

2.7. Nano-Indentation

Two of the tensile test specimens were sectioned lengthwise and prepared for nano-indentation. The instrumented indentation tests were performed using a Anton Paar NHT^3 (Anton Paar GmbH, Graz, Austria) nano indenter using a Berkovich indenter. A maximum load of 50 mN was applied at a loading and unloading rate of 60 mN/m with a dwell time of 15 s. The indentation load and deformation depth were monitored to assess the hardness across the samples. Oliver-Pharr model was used to calculate the material hardness for each indentation. Nano-indentation was performed along the centre-line of the sectioned surface up to a distance of 5 mm from the fractured surface with 100 μm intervals to identify any variation in material and potential strain hardening during tensile tests.

2.8. Scanning Electron Microscopy (SEM)

Samples from the tensile test specimens were sliced, mounted in resin and polished for metallurgical analysis. The samples were etched using a hydrochloric/sulphuric acid mixture following the method of Vander Voort [51] to reveal the microstructure. The samples were analysed using a SU3900 Scanning Electron Microscope (SEM, Hitachi HighTech, Krefeld, Germany) with an Ultimax Energy-dispersive X-ray spectroscopy (EDX, Oxford Instruments, Abingdon, UK).

3. Results

3.1. Finite Element Modelling

The peak values of maximum principal stress generally occurred at the contact between the plate and the screws (Figure 3), quite often at the screw hole immediately above the bridging span of the plate.

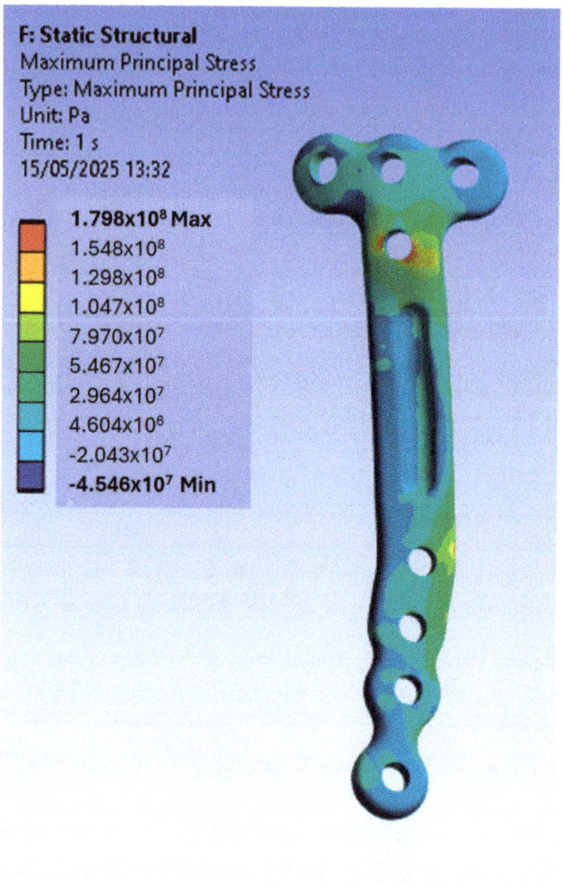

Figure 3. Representative plot of maximum principal stress plotted for a subject-specific HTO plate, peak values observed at the plate/screw interface.

The effect of the healing stage on the predicted plate stresses was substantial; the peak values of the stresses are plotted as box and whisker plots for the whole cohort for each load step in Figure 4. Over the cohort, there was considerable variation in the peak values of Von Mises, maximum and minimum principal stresses for each loading step, with a clear reduction in stress values as healing progressed from HS2 to HS4.

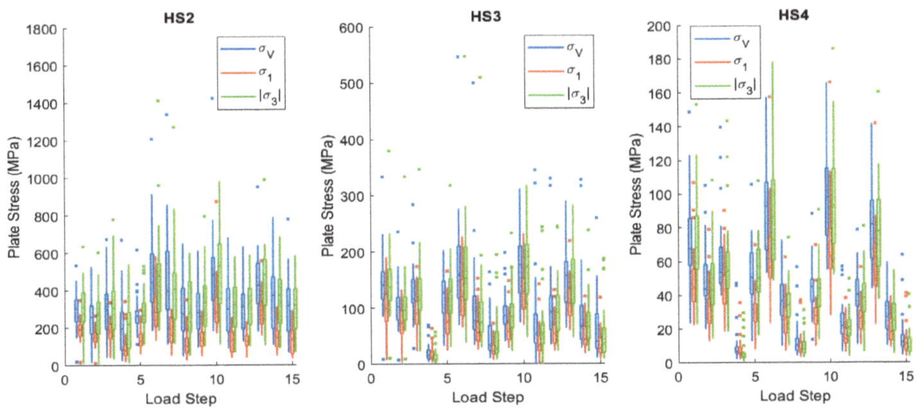

Figure 4. Box and whisker plots of stresses for whole cohort per loading step at each of the healing stages (HS) considered. Note change in *y*-axis scales as healing stage progresses from HS2 to HS4. The Von Mises stress (σ_V), maximum principal stress (σ_1) and minimum principal stress (σ_3) are plotted; note that the absolute values of σ_3 are plotted.

The peak values over the whole cohort and all loading steps for each healing stage are given in Figure 5; the peak values for maximum principal stress (σ_1) drop from 872 MPa for HS2, to 230 MPa for HS3 and 166 MPa for HS4. Peak values of Von Mises stress (σ_v) (HS2: 1423 MPa, HS3: 545 MPa, HS4: 165 MPa) were generally higher than the peak maximum principal stress, reflecting the importance of the minimum principal stress. The lowest minimum principal stress values were −1704 MPa for HS2, −546 MPa for HS3 and −186 MPa for HS4.

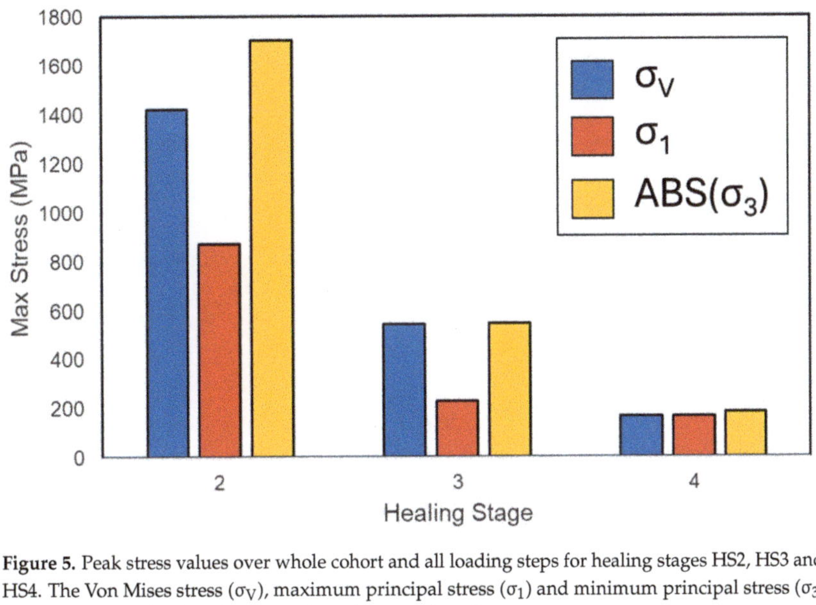

Figure 5. Peak stress values over whole cohort and all loading steps for healing stages HS2, HS3 and HS4. The Von Mises stress (σ_V), maximum principal stress (σ_1) and minimum principal stress (σ_3) are plotted; note that the absolute values of σ_3 are plotted.

3.2. Tensile Tests

The pre-test measurements of tensile test specimen working section diameter ranged from 3.45 to 3.50 mm, mean (±standard deviation or SD) value was 3.46 ± 0.02 mm (Table A1).

A representative stress-strain curve from the tensile testing is given in Figure 6. All specimens exhibited a similar behaviour, with stress increasing linearly with strain until yielding occurred, resulting in necking and ultimately failure.

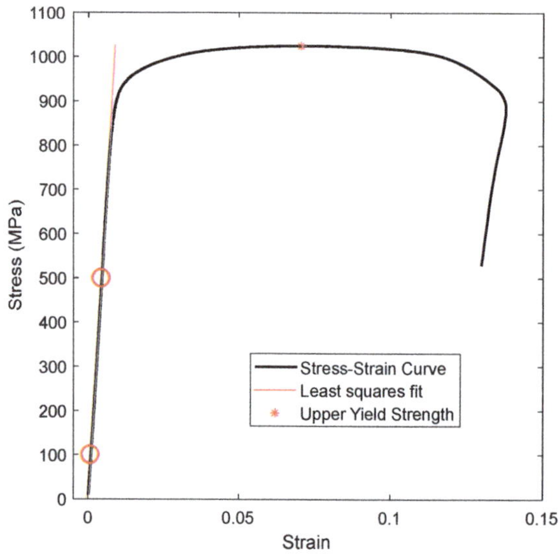

Figure 6. Strain-strain curve for representative tensile test specimen.

The mean value, over all tensile test specimens, of the peak force measured was 9518.8 ± 119.8 N, and the corresponding mean R_{eH} was 1012.5 ± 19.2 MPa (Table A2). The mean value of the 0.2% proof stress, Rp, was 911.6 ± 21.5 MPa. The mean Young's Modulus was 111.1 ± 1.7 GPa. The mean specimen minimum diameter post-test was 3.28 ± 0.03 mm, equating to a 5.10% reduction in diameter or 9.96% reduction in cross-sectional area. The average elongation at fracture across all 10 tensile specimens was 13.68 ±1.16% (Table A2).

3.3. Fatigue Tests

For the determination of the S-N curve gradient, the number of cycles to failure was more variable in the polished specimens (Group 1, Table A3). For the polished, heat-treated specimens (Group 1, Figure 7), the most probable estimate of the mean S-N relationship was:

$$\hat{\mu}_x = 6.798 - 0.00464\,y$$

where $\hat{\mu}_x$ is the log of the estimated number of cycles for a given stress y. Similarly, for the heat-treated only specimens (Group 2, Figure 7), the most probable estimate of the mean S-N relationship was:

$$\hat{\mu}_x = 7.526 - 0.00743\,y$$

The sequential progress of the staircase method for both groups is plotted in Figure 8. The fatigue strength for Group 1 (heat-treated and polished specimens) was 285.9 ± 10.3 MPa, and that for Group 2 (heat-treated only) was 226.8 ± 8.2 MPa.

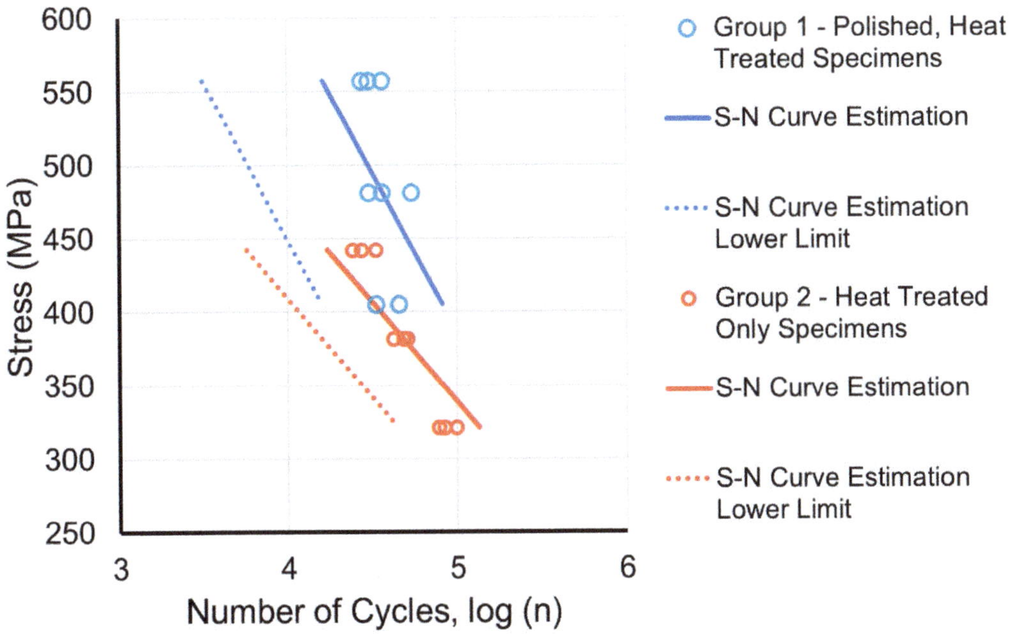

Figure 7. Initial part of S-N curve estimates for Group 1 (heat treated and polished) and Group 2 (heat treated only).

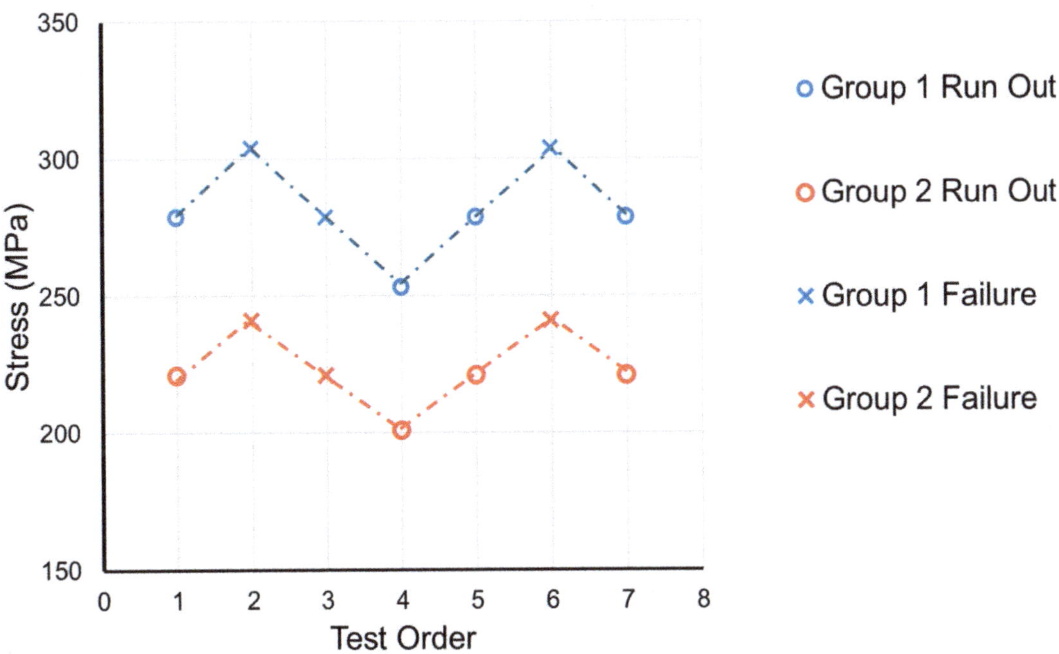

Figure 8. Staircase method results for Groups 1 (heat treated and polished) and Group 2 (heat treated only), run out occurs when failure has not occurred by 1 million cycles.

3.4. Microstructural Analysis

Figure 9 shows a representative microstructure of the LPBF Ti-6Al-4V samples after polishing and etching; all specimens had a similar appearance. The range of microstructures formed during AM has been overviewed by Tong et al. [52] and examined by Vrancken et al. [53]. The Ti–6Al–4V alloy examined in this study is an α–β alloy, where the α-phase has a hexagonal close-packed structure, and the β-phase has a body-centred cubic structure. During additive manufacturing, the heating and cooling rates can be extremely high, and can lead to the formation of a non-equilibrium martensitic α'-phase, rather than equilibrium α– and β-phases, which can increase the strength but is at the expense of reducing the ductility. A heat treatment has therefore been employed here to enable the initial martensitic α' to transform to α and β. As an example, Vranken et al. [53] reported that after a similar heat treatment of 850 °C for 2 h on SLM parts, the fully martensitic α'-phase structure of a Ti–6Al–4V was converted to a mixture of α and β, where the α-phase was present as needles. It can be seen from Figure 9 that the microstructure exhibits a similar needle-like structure as a result of the conversion of the α' martensite phase to α and β, and such a microstructure was considered to produce to best optimum mechanical properties in terms of strength and ductility [53].

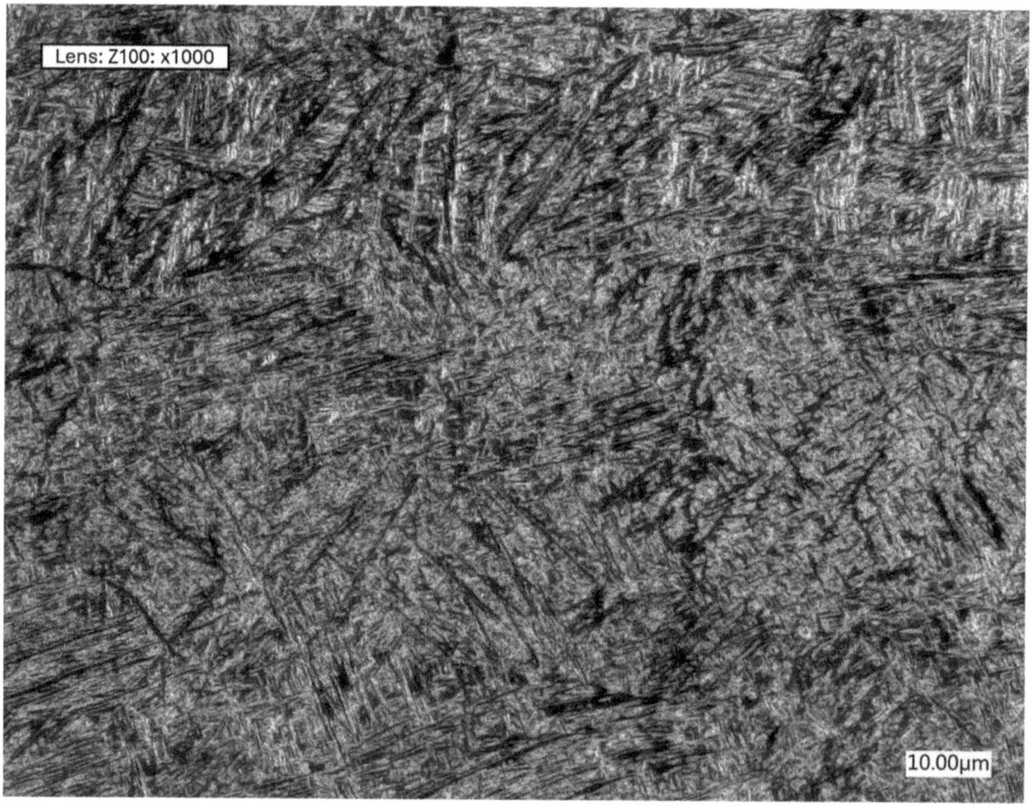

Figure 9. Microstructure of a representative LPBF Ti-6Al-4V specimen.

3.5. Nano-Indentation

Instrumented indentation was used to characterise the nano hardness of the sample as the distance increased from the fractured surface. Figure 10 illustrates the plastic deformation of the sample as the load was applied using a Berkovich indenter. The graph shows a clear distinction between the material behaviour near the fractured surface to that of the bulk material at 2.5 mm and 5 mm distance from the surface. The Oliver-Pharr method was used to calculate the nano hardness at various points from the force-plastic deformation results. The measured hardness of the tensile sample at 100 μm intervals from the fractured surface is shown in Figure 11. The sample has a bulk hardness of 429 HV. As shown in Figure 11, as the distance from the fractured surface increases, the hardness drops to reach the bulk material hardness of ~413 VH. The increased nano hardness closer to the fractured area can be attributed to strain hardening during tensile testing.

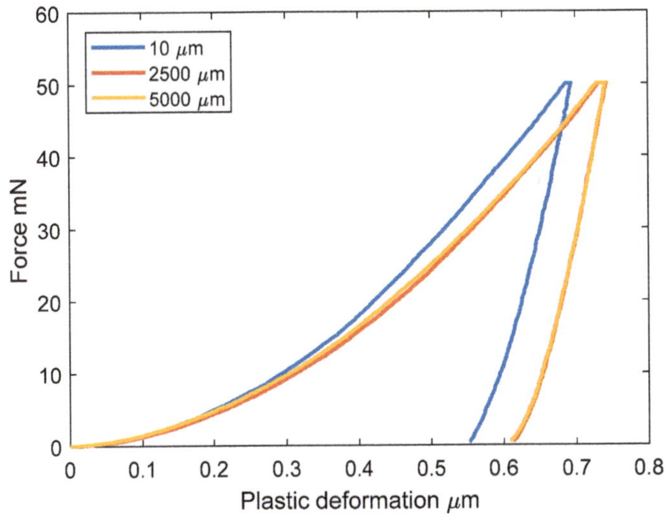

Figure 10. Force versus plastic deformation graph from instrumented indentation at 10 μm, 2500 μm and 5000 μm from the distance from the fractured surface.

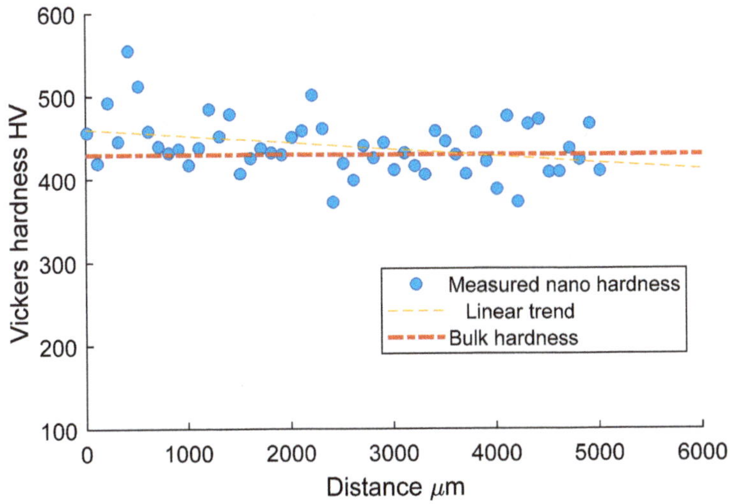

Figure 11. Measured hardness of a tensile sample at 100 μm intervals from the fractured surface.

3.6. Scanning Electron Microscopy (SEM)

The SEM micrograph of the fractured edge is shown in Figure 12. Whilst there is an indication of microstructural alignment as a result of tensile loading, the failure appears to have been initiated from the pores within the material. The presence of porosity and semi-molten powder particles in laser powder bed fusion samples is a well-known phenomenon, and typically, the density of such samples is below 99%. These can act as stress concentration points and weaken the material compared to fully dense wrought alloys. Figure 13 illustrates a high magnification micrograph of a pore.

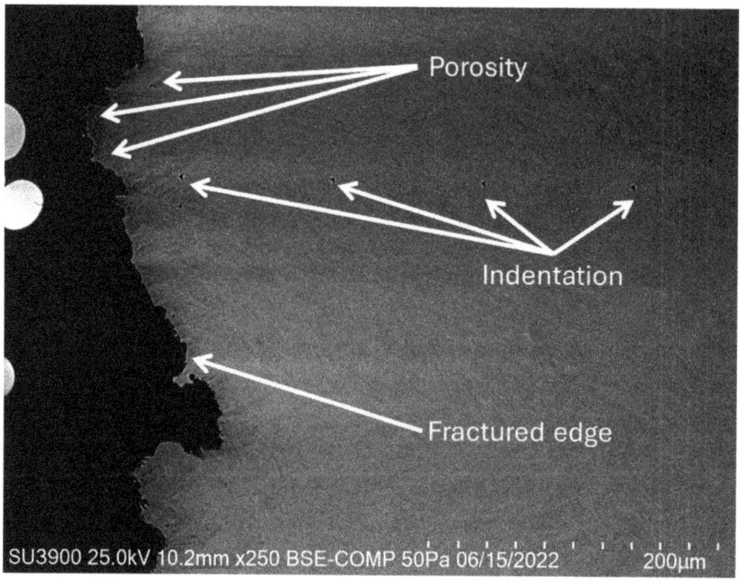

Figure 12. SEM micrograph of the fractured edge.

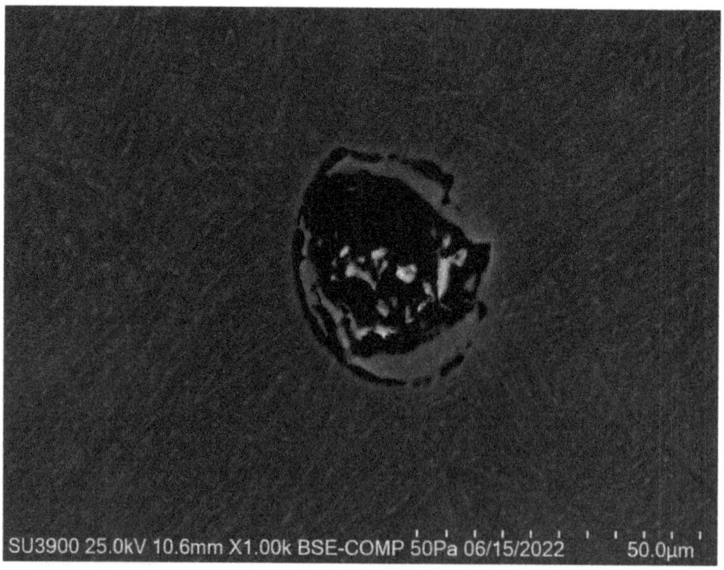

Figure 13. High magnification micrograph of a pore.

EDX analysis of the samples confirms the composition of the material mainly consisting of titanium (Ti), aluminium (Al) and vanadium (V) with trace amounts of nitrogen (N) and iron (Fe) with zirconium (Zr) present in the ceramic resin used for mounting the sample and Silicon (Si) associated with the polishing compound. The latter two elements were only detected around the edges of the sample. Figure 14 demonstrates the EDX spectrum colour maps of the main constituent elements of the sample. The analysis clearly showed that the main elements (Ti, Al and V) are evenly and randomly distributed across the sample with no areas of high concentration. Repeating the analysis at the centre of the sample, away from the edges, confirms the composition with an average of 6.3 wt% Al, 4.38 wt% V and balance Ti as shown in the spectrum analysis in Figure 15.

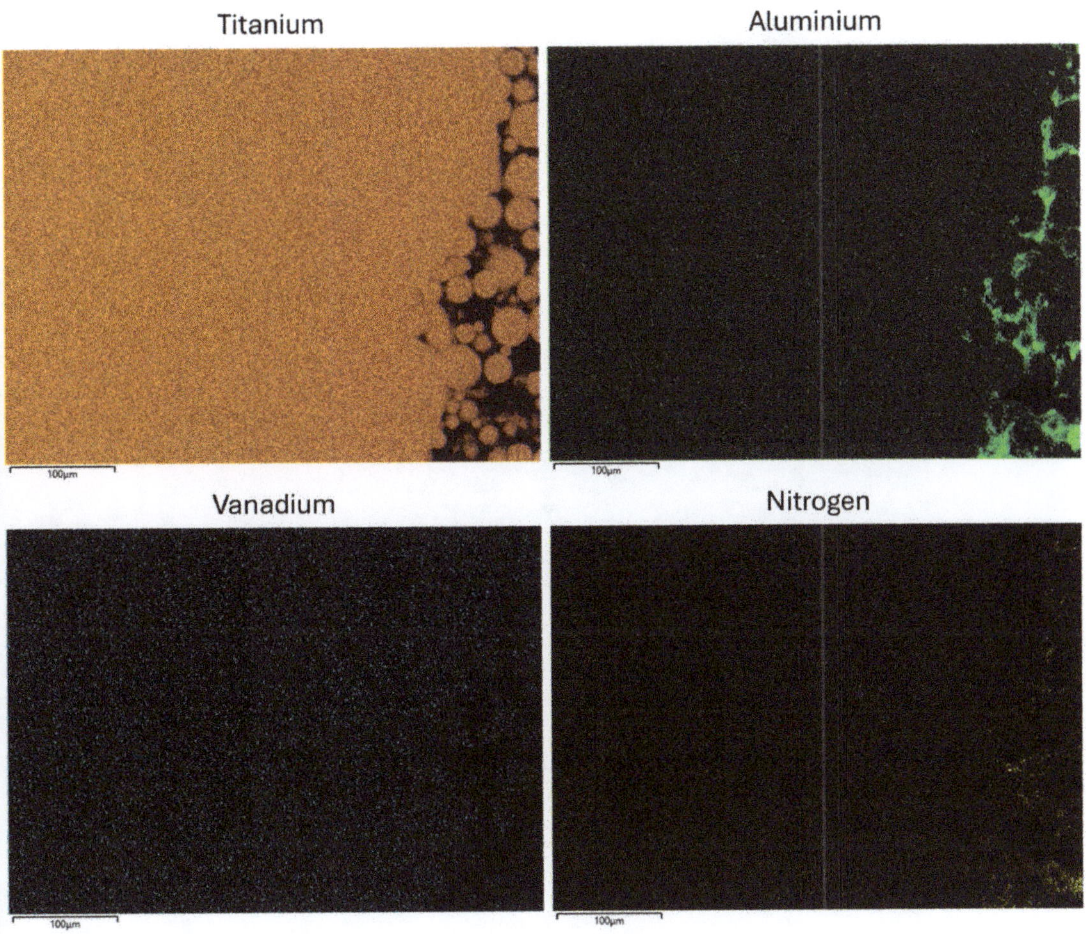

Figure 14. Distribution of the main constituent elements across the sample.

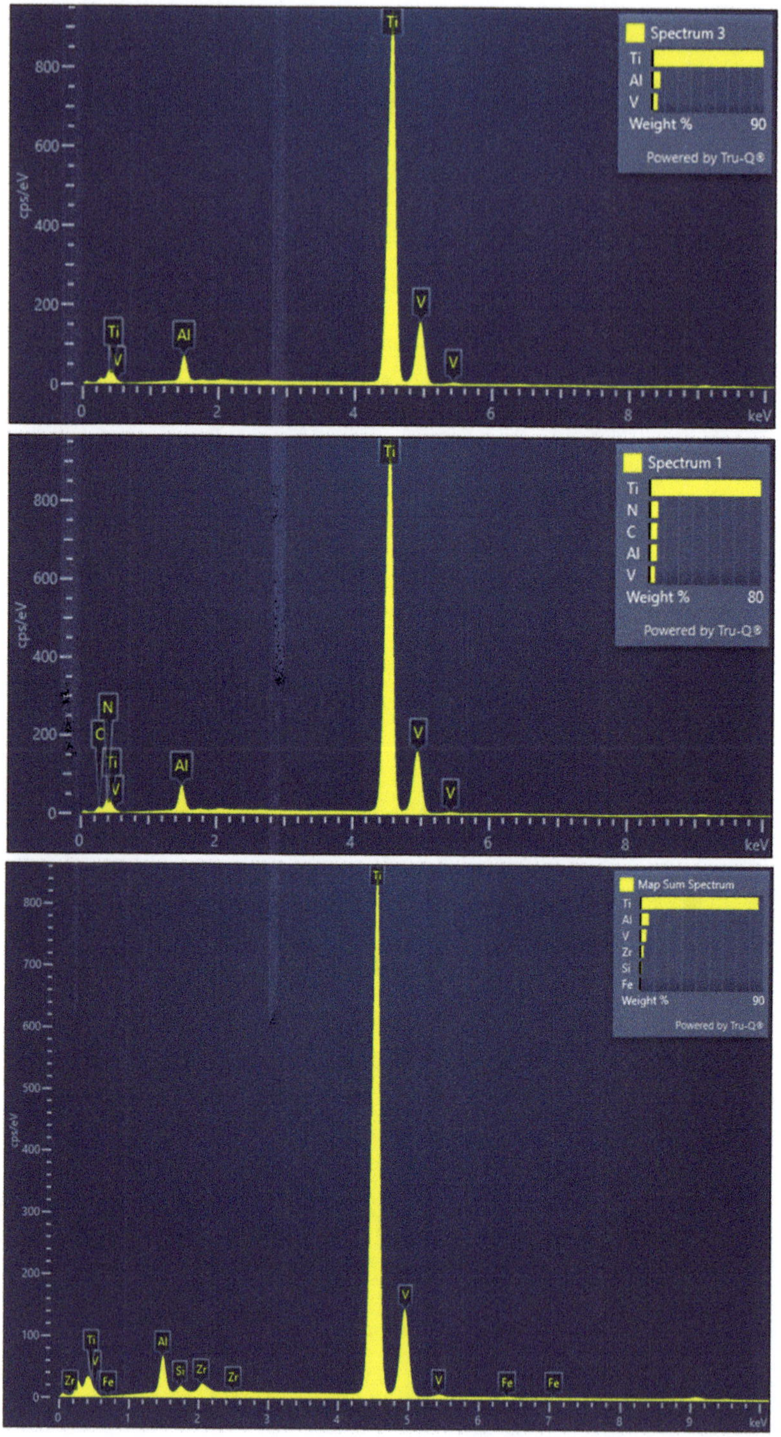

Figure 15. EDX spectrum analysis of the sample at different locations.

4. Discussion

The purpose of this study was to estimate the peak stresses in an AM osteosynthesis plate during physiological loading and establish if the mechanical properties of LPBF titanium alloy were suitable for this use case. The finite element modelling predicted peak tensile stresses of approximately 872 MPa during physiological loading at two weeks post-operatively over a cohort of 28 subjects with body weight in the range 68.8 to 121.4 kg; this peak value dropped to 230 MPa at six weeks and 166 MPa at twelve weeks post-operatively.

A series of tensile and fatigue tests were performed on additively manufactured specimens. Comparing the tensile test results with the standard industry values [46,54] and published values [55], the findings were within the normal range for the alloy. For additive manufacture, the vertical direction has lower values compared to the horizontal direction for Young's Modulus and ultimate tensile strength [46]. This study found that the Young's modulus (111.1 GPa) was at the lower end of the additive machine manufacturer's values. The percentage elongation at breakage of the titanium alloy (13.68%) was within the reported range. The yield strength was found to be 911.6 $\pm$ 21.5 MPa, similar to that reported by Vrancken et al. [53], which was lower than the value quoted by the additive machine manufacturer but within the range quoted for the material [54]. The values for yield strength and upper yield strength are comparable to those quoted for yield strength and ultimate tensile strength by Alcisto et al. [55] for wrought titanium alloy, and so are the percentage elongation values. Alcisto et al. reported that whilst yield strength and ultimate tensile strength of their additively manufactured material was similar to wrought material, the percentage elongation was substantially lower. A possible explanation for the improved percentage elongation could be the combination of smaller grain size powder and higher laser power in more recent additive manufacturing machines.

The microscopy analysis of the microstructure showed the appearance typical of LPBF Ti-6Al-4V parts that have been heat-treated [56]. The indentation tests confirmed strain hardening behaviour during tensile testing. SEM analysis showed the presence of porosity, as is common for laser power bed fusion parts, with EDX analysis showing the expected composition of the titanium alloy. LPBF Ti-6Al-4V samples can typically reach up to 99% density. The presence of pores can reduce the mechanical and fatigue strength of the samples through reduction of cross cross-sectional area and/or introducing high stress concentration areas [57]. Pessard et al. [58] identified a critical defect size of 30 μm. They noted that pores with $\sqrt{area}$ larger than this threshold negatively impact the fatigue strength of LPBF Ti-6Al-4V alloy. Porosity can be controlled by selecting appropriate AM processing parameters, i.e., scanning speed and hatch distance and through postprocessing such as high-pressure isostatic pressing [59].

The LPBF Ti-6Al-4V material exceeded the minimum requirements of ISO 5832-3:2019 Implants for surgery—Metallic materials. Part 3: Wrought titanium 6-aluminium 4-vanadium alloy for proof strength (800 MPa), ultimate tensile strength (900 MPa) and percentage elongation (10%).

Patients who have received orthopaedic treatment walk approximately 1.9 million cycles a year [60]; however, during the early recovery period, partial weight bearing is normally prescribed, which significantly reduces peak ground reaction forces [61]. For lower limb osteotomies, healing (defined as bone union) occurs in 4.5 $\pm$ 1.5 months [62]. Once healing has occurred, the majority of the load is transmitted through the bone and the forces passing through the osteosynthesis device are significantly reduced. The fatigue testing reported here demonstrated the effects of surface finish, heat-treated and polished specimens had a fatigue strength of approximately 286 MPa, whilst that for specimens which were only heat-treated was approximately 227 MPa. This finding is supported by those of Hu et al. [34], who reported surface defects had a greater effect on fatigue performance than internal defects. Even without polishing, additively manufactured osteosynthesis devices exceed the functional fatigue requirements for this class of device. The finite element simulations performed in the current study used full weight-bearing forces and joint reactions. Typically, osteotomy patients are told to refrain from full weight-bearing for six weeks. Although, even when early weight-bearing is encouraged, it is likely to occur between two and five weeks following surgery [63]. Therefore, the stresses in the plate at Healing Stage 2 represent a worst-case scenario, and even given this worst-case scenario, the stress in the plate was comfortably below the yield strength. The maximum load at six weeks, when patients are more likely to be fully weight bearing, was also well below the fatigue limit of the heat-treated and polished material. Three months post-operatively, the peak loads in additively manufactured osteotomy plates were well below 180 MPa. For osteosynthesis applications using additive manufacturing with Ti-Al6-V4, we would recommend that components be heat-treated and polished.

A limitation of this study is that only specimens from one particular type of AM machine, which used the SLM process, were tested. The specimens were manually finished or polished where appropriate, which introduces variability into the manufacturing process; however, this is representative of custom-made medical devices, where each component/system is unique. The plate is custom, and therefore, the shape changes for every single patient; changes in geometry can significantly influence stresses. The finite element modelling was performed for a cohort of 28 subjects with knee OA, and therefore covered a range of geometry variations.

5. Conclusions

The tensile and fatigue properties of heat-treated LPBF Ti-6Al-4V are sufficient to meet the mechanical requirements of osteosynthesis plates. Whilst 99% density can be achieved in laser powder-bed fusion (LPBF) additive manufacturing, the presence of porosity can impact the mechanical and fatigue performance of the additively manufactured samples and act as stress concentration areas. Heat treatment and polishing of components are recommended for osteosynthesis applications.

Author Contributions: Conceptualization, A.R.M., A.C.L. and H.S.G.; methodology, A.R.M., M.B., A.S., C.R.B. and H.S.G.; software, A.R.M., M.B. and H.S.G.; validation, A.S., C.R.B. and H.S.G.; formal analysis, A.R.M., M.B., A.S. and H.S.G.; investigation, A.R.M., M.B., A.S. and H.S.G.; resources, A.S., C.R.B., A.C.L. and H.S.G.; data curation, A.S., C.R.B. and H.S.G.; writing—original draft preparation, A.R.M., A.S., C.R.B. and H.S.G.; writing—review and editing, A.R.M., M.B., A.S., A.C.L., C.R.B. and H.S.G.; visualization, A.R.M., A.S., M.B. and H.S.G.; supervision, A.R.M., A.C.L. and H.S.G.; project administration, H.S.G.; funding A.C.L. quisition, H.S.G. All authors have read and agreed to the published version of the manuscript.

Funding: The lead author was funded by Arthritis Research UK, grant 21495, and Versus Arthritis, grant 22262. Additional funding came from the SBRI Healthcare Programme Phase 2 award, grant 32619930. MB was funded by an Innovate KTP award. The funders played no role in the work performed and the preparation of the manuscript. The authors acknowledge the support of Orthoscape, 3D Metal Printing Ltd and Renishaw plc, who supplied components and specimens.

Data Availability Statement: The data presented in this study are available on request from the corresponding author due to delays in depositing data with Institutional repository.

Conflicts of Interest: A.R.M, A.C.L. and H.S.G. are named as inventors on related patent GB2551533, held jointly by the University of Bath, Royal Devon & Exeter NHS Foundation Trust and 3D EMS Ltd. A.C.L. and M.B. are employees of Orthoscape, 3D Metal Printing Ltd. The other authors declare no conflict of interest.

Appendix A

Table A1. Reduction in working section diameter as a result of testing for each of the tensile test specimens.

Specimen	Pre Average Diameter (mm)	Post Average Diameter (mm)	Difference (mm)	Percentage Difference (%)
1	3.46	3.28	−0.17	−5.01
2	3.47	3.25	−0.22	−6.35
3	3.45	3.27	−0.18	−5.31
4	3.50	3.35	−0.15	−4.29
5	3.46	3.31	−0.15	−4.43
6	3.46	3.27	−0.19	−5.49
7	3.46	3.25	−0.21	−6.17
8	3.44	3.24	−0.20	−5.72
9	3.45	3.30	−0.14	−4.16
10	3.45	3.31	−0.14	−4.06
Average	3.46	3.28	−0.18	−5.10

Table A2. Peak force, peak stress and Young's Modulus from the tensile testing.

Specimen	Peak Force (MPa)	Peak Stress (MPa)	Youngs Modulus, E (MPa)	Initial Gauge Length (mm)	Maximum Extension (mm)	Maximum Extension (%)
1	9551	1015	112,679	19.90	2.74	13.78
2	9563	1011	107,801	20.50	2.92	14.23
3	9373	1003	110,616	21.49	3.09	14.38
4	9353	972	112,737	21.64	2.48	11.47
5	9382	998	111,863	20.91	2.76	13.20
6	9624	1024	111,321	21.50	2.54	11.80
7	9423	1002	108,416	20.75	3.20	15.41
8	9696	1043	113,196	21.02	3.04	14.47
9	9566	1023	110,096	20.87	3.00	14.38
10	9656	1033	112,079	21.55	2.94	13.66
Mean	9519	1012	111,080	21.01	2.87	13.68
SD	120	19	1746	0.52	0.22	1.16

Table A3. Fatigue test results for the slope of the S-N curve, showing cycles to failure for a given applied stress, Group 1 heat treated and polished, Group 2 heat treated only.

		Number of Cycles to Failure				
		Repetition Number				
Group	**Stress (MPa)**	**1**	**2**	**3**	**Mean**	**SD**
1	557	27,351	36,468	30,390	31,403	4642
	481	36,468	30,390	54,702	40,520	12,652
	405	45,585	45,585	33,429	41,533	7018
2	442	33,429	27,351	24,312	28,364	4642
	382	48,624	51,663	42,546	47,611	4642
	321	100,287	85,092	79,014	88,131	10,957

Finite Element Modelling: Figure A1a shows an exemplar of the overall model, including the proximal tibia, the plate and the screws. A typical physiological set of loads and the boundary conditions imposed at the distal aspect of the proximal tibia are shown in Figure A1b. A representative mesh is given in Figure A1c, showing the differences in mesh density between the bone and the plate, and the refinement where contact between plate and screws was modelled. The locations at which contact was modelled between the plates and the screws are indicated in Figure A1d.

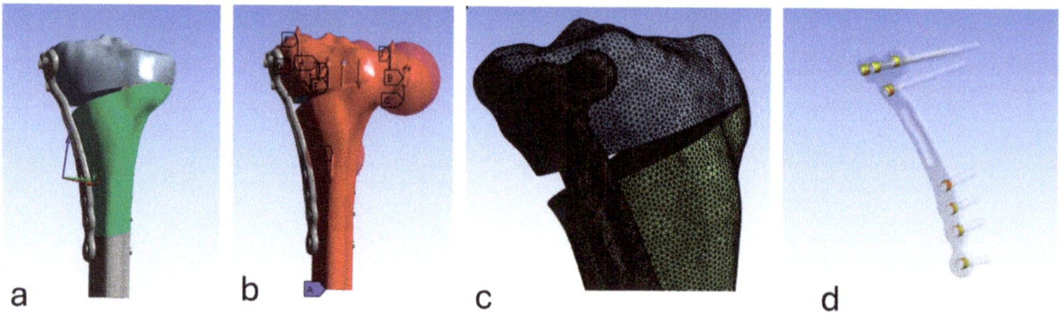

Figure A1. Details of the finite element modelling, (**a**) exemplar overall model arrangement, (**b**) typical physiological loads (red) and boundary condition at distal aspect of proximal tibia, (**c**) details of meshing, showing differences in mesh density between bone, plate and screw as well as local refinement around contact locations, (**d**) locations at which contact was modelled between plate and screws.

References

1. Auricchio, F.; Marconi, S. 3D printing: Clinical applications in orthopaedics and traumatology. *EFORT Open Rev.* **2016**, *1*, 121–127. [CrossRef] [PubMed]
2. Mulford, J.S.; Babazadeh, S.; Mackay, N. Three-dimensional printing in orthopaedic surgery: Review of current and future applications. *ANZ J. Surg.* **2016**, *86*, 648–653. [CrossRef] [PubMed]
3. Mumith, A.; Thomas, M.; Shah, Z.; Coathup, M.; Blunn, G. Additive manufacturing: Current concepts, future trends. *Bone Jt. J.* **2018**, *100-B*, 455–460. [CrossRef] [PubMed]
4. Zhong, S.; Shi, Q.; Van Dessel, J.; Gu, Y.; Sun, Y.; Yang, S.; Constantinus, P. Biomechanical validation of structural optimized patient-specific mandibular reconstruction plate orienting additive manufacturing. *Comput. Methods Programs Biomed.* **2022**, *224*, 107023. [CrossRef]
5. Robinson, D.L.; Bucknill, A.; Ferragina, A.; Campbell, C.; Lee, P.V.S. Fixation of pelvic acetabular fractures using 3D-printed fracture plates: A cadaver study. *J. Orthop. Surg. Res.* **2023**, *18*, 360. [CrossRef]
6. Sun, H.; Hu, C.; Zhou, C.; Wu, L.; Sun, J.; Zhou, X.; Xing, F.; Long, C.; Kong, Q.; Liang, J.; et al. 3D printing of calcium phosphate scaffolds with controlled release of antibacterial functions for jaw bone repair. *Mater. Des.* **2020**, *189*, 108540. [CrossRef]
7. Wong, K.C.; Kumta, S.M.; Geel, N.V.; Demol, J. One-step reconstruction with a 3D-printed, biomechanically evaluated custom implant after complex pelvic tumor resection. *Comput. Aided Surg.* **2015**, *20*, 14–23. [CrossRef]
8. Wong, K.C. 3D-printed patient-specific applications in orthopedics. *Orthop. Res. Rev.* **2016**, *8*, 57–66. [CrossRef]
9. Perticarini, L.; Zanon, G.; Rossi, S.M.; Benazzo, F.M. Clinical and radiographic outcomes of a trabecular titanium acetabular component in hip arthroplasty: Results at minimum 5 years follow-up. *BMC Musculoskelet. Disord.* **2015**, *16*, 375. [CrossRef]
10. De Meo, F.; Cacciola, G.; Bellotti, V.; Bruschetta, A.; Cavaliere, P. Trabecular Titanium acetabular cups in hip revision surgery: Mid-term clinical and radiological outcomes. *Hip Int.* **2018**, *28*, 61–65. [CrossRef]
11. Faizan, A.; Bhowmik-Stoker, M.; Alipit, V.; Kirk, A.E.; Krebs, V.E.; Harwin, S.F.; Meneghini, R.M. Development and Verification of Novel Porous Titanium Metaphyseal Cones for Revision Total Knee Arthroplasty. *J. Arthroplast.* **2017**, *32*, 1946–1953. [CrossRef]
12. Li, Z.; Luo, Y.; Lu, M.; Wang, Y.; Gong, T.; Hu, X.; He, X.; Zhou, Y.; Min, L.; Tu, C. 3D-printed Personalized Porous Acetabular Component to Reconstruct Extensive Acetabular Bone Defects in Primary Hip Arthroplasty. *Orthop. Surg.* **2024**, *16*, 1642–1647. [CrossRef] [PubMed]
13. Geetha, M.; Singh, A.K.; Asokamani, R.; Gogia, A.K. Ti based biomaterials, the ultimate choice for orthopaedic implants—A review. *Progress. Mater. Sci.* **2009**, *54*, 397–425. [CrossRef]
14. *ASTM F136-13*; Standard Specification for Wrought Titanium-6Aluminum-4Vanadium ELI (Extra Low Interstitial) Alloy for Surgical Implant Applications (UNS R56401). ASTM International: West Conshohocken, PA, USA, 2013. [CrossRef]
15. *ASTM F3001-14*; Standard Specification for Additive Manufacturing Titanium-6 Aluminum-4 Vanadium ELI (Extra Low Interstitial) with Powder Bed Fusion. ASTM International: West Conshohocken, PA, USA, 2014. [CrossRef]

16. *ISO 5832-3:2016*; Implants for Surgery—Metallic Materials. Part 3: Wrought Titanium 6-aluminium 4-vanadium Alloy. ISO: Geneva, Switzerland, 2016.
17. Hollander, D.A.; Wirtz, T.; von Walter, M.; Linker, R.; Schultheis, A.; Paar, O. Development of IndividualThree-Dimensional Bone Substitutes Using "Selective LaserMelting". *Eur. J. Trauma.* **2003**, *29*, 228–234. [CrossRef]
18. Haslauer, C.M.; Springer, J.C.; Harrysson, O.L.; Loboa, E.G.; Monteiro-Riviere, N.A.; Marcellin-Little, D.J. In vitro biocompatibility of titanium alloy discs made using direct metal fabrication. *Med. Eng. Phys.* **2010**, *32*, 645–652. [CrossRef]
19. Sidambe, A.T. Biocompatibility of Advanced Manufactured Titanium Implants—A Review. *Materials* **2014**, *7*, 8168–8188. [CrossRef]
20. Lin, X.; Xiao, X.; Wang, Y.; Gu, C.; Wang, C.; Chen, J.; Liu, H.; Luo, J.; Li, T.; Wang, D.; et al. Biocompatibility of Bespoke 3D-Printed Titanium Alloy Plates for Treating Acetabular Fractures. *Biomed. Res. Int.* **2018**, *2018*, 2053486. [CrossRef] [PubMed]
21. Liang, H.; Yang, Y.; Xie, D.; Li, L.; Mao, N.; Wang, C.; Tian, Z.; Jiang, Q.; Shen, L. Trabecular-like Ti-6Al-4V scaffolds for orthopedic: Fabrication by selective laser melting and in vitro biocompatibility. *J. Mater. Sci. Technol.* **2019**, *35*, 1284–1297. [CrossRef]
22. Tuomi, J.T.; Björkstrand, R.V.; Pernu, M.L.; Salmi, M.V.J.; Huotilainen, E.I.; Wolff, J.E.H.; Vallittu, P.K.; Mäkitie, A.A. In vitro cytotoxicity and surface topography evaluation of additive manufacturing titanium implant materials. *J. Mater. Sci. Mater. Med.* **2017**, *28*, 53. [CrossRef]
23. Jennison, T.; Brinsden, M. Fracture admission trends in England over a ten-year period. *Ann. R. Coll. Surg. Engl.* **2019**, *101*, 208–214. [CrossRef]
24. Banerjee, S.; Kulesha, G.; Kester, M.; Mont, M.A. Emerging technologies in arthroplasty: Additive manufacturing. *J. Knee Surg.* **2014**, *27*, 185–191. [CrossRef]
25. Chen, X.; Possel, J.K.; Wacongne, C.; van Ham, A.F.; Klink, P.C.; Roelfsema, P.R. 3D printing and modelling of customized implants and surgical guides for non-human primates. *J. Neurosci. Methods* **2017**, *286*, 38–55. [CrossRef]
26. Bayliss, L.E.; Culliford, D.; Monk, A.P.; Glyn-Jones, S.; Prieto-Alhambra, D.; Judge, A.; Cooper, C.; Carr, A.J.; Arden, N.K.; Beard, D.J.; et al. The effect of patient age at intervention on risk of implant revision after total replacement of the hip or knee: A population-based cohort study. *Lancet* **2017**, *389*, 1424–1430. [CrossRef] [PubMed]
27. Wainwright, C.; Theis, J.C.; Garneti, N.; Melloh, M. Age at hip or knee joint replacement surgery predicts likelihood of revision surgery. *J. Bone Jt. Surg. Br.* **2011**, *93*, 1411–1415. [CrossRef] [PubMed]
28. Murray, R.; Winkler, P.W.; Shaikh, H.S.; Musahl, V. High Tibial Osteotomy for Varus Deformity of the Knee. *J. Am. Acad. Orthop. Surg. Glob. Res. Rev.* **2021**, *5*, e1. [CrossRef]
29. Bergmann, G.; Bender, A.; Dymke, J.; Duda, G.; Damm, P. Standardized Loads Acting in Hip Implants. *PLoS ONE* **2016**, *11*, e0155612. [CrossRef] [PubMed]
30. Bergmann, G.; Bender, A.; Graichen, F.; Dymke, J.; Rohlmann, A.; Trepczynski, A.; Heller, M.O.; Kutzner, I. Standardized loads acting in knee implants. *PLoS ONE* **2014**, *9*, e86035. [CrossRef]
31. Schorler, H.; Capanni, F.; Gaashan, M.; Wendlandt, R.; Jurgens, C.; Schulz, A.P. Bone plates for osteosynthesis - a systematic review of test methods and parameters for biomechanical testing. *Biomed. Tech.* **2017**, *62*, 235–243. [CrossRef]
32. MacLeod, A.R.; Pankaj, P. Pre-operative planning for fracture fixation using locking plates: Device configuration and other considerations. *Injury* **2018**, *49* (Suppl. S1), S12–S18. [CrossRef]
33. MacLeod, A.R.; Serrancoli, G.; Fregly, B.J.; Toms, A.D.; Gill, H.S. The effect of plate design, bridging span, and fracture healing on the performance of high tibial osteotomy plates: An experimental and finite element study. *Bone Jt. Res.* **2018**, *7*, 639–649. [CrossRef]
34. Hu, Y.N.; Wu, S.C.; Withers, P.J.; Zhang, J.; Bao, H.Y.X.; Fu, Y.N.; Kang, G.Z. The effect of manufacturing defects on the fatigue life of selective laser melted Ti-6Al-4V structures. *Mater. Des.* **2020**, *192*, 108708. [CrossRef]
35. Chen, G.; Schmutz, B.; Wullschleger, M.; Pearcy, M.J.; Schuetz, M.A. Computational investigations of mechanical failures of internal plate fixation. *Proc. Inst. Mech. Eng. H.* **2010**, *224*, 119–126. [CrossRef] [PubMed]
36. Nica, M.; Cretu, B.; Ene, D.; Antoniac, I.; Gheorghita, D.; Ene, R. Failure Analysis of Retrieved Osteosynthesis Implants. *Materials* **2020**, *13*, 1201. [CrossRef]
37. MacLeod, A.; Peckham, N.; Serrancolí, G.; Rombach, I.; Hourigan, P.; Mandalia, V.I.; Toms, A.D.; Fregly, B.J.; Gill, H.S. Personalised high tibial osteotomy has mechanical safety equivalent to generic device in a case-control in silico clinical trial. *Commun. Med.* **2021**, *1*, 6. [CrossRef]
38. Mohammed, M.M. Fundamental Drivers of Metal Fatigue Failure and Performance. *Multidiscip. Mater. Chron.* **2024**, *1*, 49–61. [CrossRef]
39. Sabzevari, S.; Ebrahimpour, A.; Roudi, M.K.; Kachooei, A.R. High Tibial Osteotomy: A Systematic Review and Current Concept. *Arch. Bone Jt. Surg.* **2016**, *4*, 204–212.
40. Taddei, F. Bonemat. Available online: https://bonemat.ior.it/en (accessed on 14 May 2025).
41. Taddei, F.; Pancanti, A.; Viceconti, M. An improved method for the automatic mapping of computed tomography numbers onto finite element models. *Med. Eng. Phys.* **2004**, *26*, 61–69. [CrossRef] [PubMed]

42. Isaksson, H.; Wilson, W.; van Donkelaar, C.C.; Huiskes, R.; Ito, K. Comparison of biophysical stimuli for mechano-regulation of tissue differentiation during fracture healing. *J. Biomech.* **2006**, *39*, 1507–1516. [CrossRef]
43. Steiner, M.; Claes, L.; Ignatius, A.; Niemeyer, F.; Simon, U.; Wehner, T. Prediction of fracture healing under axial loading, shear loading and bending is possible using distortional and dilatational strains as determining mechanical stimuli. *J. R. Soc. Interface* **2013**, *10*, 20130389. [CrossRef]
44. MacLeod, A.R.; Pankaj, P.; Simpson, A.H. Does screw-bone interface modelling matter in finite element analyses? *J. Biomech* **2012**, *45*, 1712–1716. [CrossRef]
45. Serrancoli, G.; Kinney, A.L.; Fregly, B.J.; Font-Llagunes, J.M. Neuromusculoskeletal Model Calibration Significantly Affects Predicted Knee Contact Forces for Walking. *J. Biomech. Eng.* **2016**, *138*, 0810011. [CrossRef] [PubMed]
46. Renishaw plc. *Datasheet: Ti6Al4V ELI-0406 Powder for Additive Manufacturing*; Renishaw plc: Wotton-under-Edge, UK, 2017.
47. Xiao, Z.; Chen, C.; Zhu, H.; Hu, Z.; Nagarajan, B.; Guo, L.; Zeng, X. Study of residual stress in selective laser melting of Ti6Al4V *Mater. Des.* **2020**, *193*, 108846. [CrossRef]
48. *ISO 6892-1:2019*; Metallic materials—Tensile testing. Part 1: Method. of test at room temperature. ISO: Geneva, Switzerland, 2019
49. Guilherme, A.S.; Henriques, G.E.; Zavanelli, R.A.; Mesquita, M.F. Surface roughness and fatigue performance of commercially pure titanium and Ti-6Al-4V alloy after different polishing protocols. *J. Prosthet. Dent.* **2005**, *93*, 378–385. [CrossRef] [PubMed]
50. *ISO 12107:2012*; Metallic materials—Fatigue testing. Statistical planning and analysis of data. ISO: Geneva, Switzerland, 2012.
51. Vander Voort, G.F. *Metallography: Principles and Practice*; ASM International: Materials Park, OH, USA, 1984.
52. Tong, J.; Bowen, C.R.; Persson, J.; Plummer, A. Mechanical properties of titanium-based Ti–6Al–4V alloys manufactured by powder bed additive manufacture. *Mater. Sci. Technol.* **2017**, *33*, 138–148. [CrossRef]
53. Vrancken, B.; Thijs, L.; Kruth, J.-P.; Van Humbeeck, J. Heat treatment of Ti6Al4V produced by Selective Laser Melting: Microstructure and mechanical properties. *J. Alloys Compd.* **2012**, *541*, 177–185. [CrossRef]
54. Titanium Metals Corporation. *Datasheet: TIMETAL 6-4, 6-4 ELI & 6-4-.1R U Medium to High Strength General-Purpose Alloys*; Titanium Metals Corporation: Warrensville Heights, OH, USA, 2000.
55. Alcisto, J.; Enriquez, A.; Garcia, H.; Hinkson, S.; Steelman, T.; Silverman, E.; Valdovino, P.; Gigerenzer, H.; Foyos, J.; Ogren, J.; et al. Tensile Properties and Microstructures of Laser-Formed Ti-6Al-4V. *J. Mater. Eng. Perform.* **2011**, *20*, 203–212. [CrossRef]
56. Zhang, S.; Lin, X.; Chen, J.; Huang, W. Heat-treated microstructure and mechanical properties of laser solid forming Ti-6Al-4V alloy. *Rare Met.* **2009**, *28*, 537–544. [CrossRef]
57. Kaletsch, A.; Qin, S.; Herzog, S.; Broeckmann, C. Influence of high initial porosity introduced by laser powder bed fusion on the fatigue strength of Inconel 718 after post-processing with hot isostatic pressing. *Addit. Manuf.* **2021**, *47*, 102331. [CrossRef]
58. Pessard, E.; Lavialle, M.; Laheurte, P.; Didier, P.; Brochu, M. High-cycle fatigue behavior of a laser powder bed fusion additive manufactured Ti-6Al-4V titanium: Effect of pores and tested volume size. *Int. J. Fatigue* **2021**, *149*, 106206. [CrossRef]
59. Du, C.; Zhao, Y.; Jiang, J.; Wang, Q.; Wang, H.; Li, N.; Sun, J. Pore defects in Laser Powder Bed Fusion: Formation mechanism, control method, and perspectives. *J. Alloys Compd.* **2023**, *944*, 169215. [CrossRef]
60. Silva, M.; Shepherd, E.F.; Jackson, W.O.; Dorey, F.J.; Schmalzried, T.P. Average patient walking activity approaches 2 million cycles per year: Pedometers under-record walking activity. *J. Arthroplast.* **2002**, *17*, 693–697. [CrossRef] [PubMed]
61. Ganse, B.; Yang, P.F.; Gardlo, J.; Gauger, P.; Kriechbaumer, A.; Pape, H.C.; Koy, T.; Muller, L.P.; Rittweger, J. Partial weight bearing of the tibia. *Injury* **2016**, *47*, 1777–1782. [CrossRef] [PubMed]
62. Takahara, Y.; Furumatsu, T.; Nakashima, H.; Itani, S.; Nakamura, M.; Uchida, Y.; Kato, H.; Tsujimura, Y.; Iwasaki, Y.; Ochi, N. Time to Bone Union after Hybrid Closed-Wedge High Tibial Osteotomy. *Acta Med. Okayama* **2019**, *73*, 511–516. [CrossRef]
63. Brinkman, J.M.; Luites, J.W.; Wymenga, A.B.; van Heerwaarden, R.J. Early full weight bearing is safe in open-wedge high tibial osteotomy. *Acta Orthop.* **2010**, *81*, 193–198. [CrossRef] [PubMed]

Journal of
Manufacturing and
Materials Processing

MDPI

Article

Exploring the Mechanical Properties of Bioprinted Multi-Layered Polyvinyl Alcohol Cryogel for Vascular Applications

Argyro Panieraki [1,*], Nasim Mahmoodi [1], Carl Anthony [1], Rosemary J. Dyson [2] and Lauren E. J. Thomas-Seale [1]

1 School of Engineering, University of Birmingham, Birmingham B15 2TT, UK; n.mahmoodi@bham.ac.uk (N.M.); c.j.anthony@bham.ac.uk (C.A.); l.e.j.thomas-seale@bham.ac.uk (L.E.J.T.-S.)
2 School of Mathematics, University of Birmingham, Birmingham B15 2TT, UK; r.j.dyson@bham.ac.uk
* Correspondence: axp308@student.bham.ac.uk

Abstract: Polyvinyl alcohol cryogels (PVA-C) are promising materials for vascular tissue engineering due to their biocompatibility, hydrophilicity, and tuneable mechanical properties. This study investigates the mechanical performance of multi-layered PVA-C constructs fabricated via sub-zero extrusion-based three-dimensional (3D) bioprinting. Samples with two, four, and six alternating layers were evaluated to assess the effect of layered architecture on elastic and viscoelastic behaviour. Uniaxial tensile testing revealed that increasing the number of layers led to a moderate reduction in stiffness; for instance, at 20% strain, six-layered constructs showed a significantly lower ($p < 0.05$) Young's modulus (36.7 ± 2.5 kPa) compared to two-layered ones (47.3 ± 3.1 kPa). Stress–strain curves exhibited nonlinear characteristics, better captured by quadratic (as opposed to linear) fitting, within the tested strain range (≤40%). Dynamic mechanical analysis demonstrated a frequency-independent storage modulus (E') across 1–10 Hz, with subtle variations in viscoelastic response linked to the number of layers. Visual inspection confirmed improved print fidelity and hydration retention in thicker constructs. These findings demonstrate that a multi-layered design influences the mechanical profile of PVA-C and suggests potential for functionally graded design strategies to enhance compliance matching and mimic the biomechanics of native vessels in small-diameter vascular grafts.

Keywords: polyvinyl alcohol cryogel; 3D bioprinting; multi-layered hydrogels; vascular grafts; mechanical properties; dynamic mechanical analysis; viscoelasticity; functionally graded materials

1. Introduction

Cardiovascular diseases remain the leading cause of mortality worldwide, with coronary artery disease being a significant contributor that often necessitates surgical interventions such as vascular grafting [1]. Synthetic vascular grafts must closely mimic the biomechanical properties of native arteries to ensure long-term functionality and reduce complications such as restenosis and thrombosis [2]. Arterial tissues exhibit a sophisticated balance of strength, stiffness, and toughness, achieved through their hierarchical structure, consisting of distinct layers with unique mechanical and biological functions [3]. Crucially, native arteries display anisotropic mechanical behaviour, primarily due to the alignment of collagen fibres and elastin within their layered architecture, which enables vessels to accommodate pulsatile blood flow while maintaining structural integrity [4].

The arterial wall consists of three distinct layers—intima, media, and adventitia—each contributing to the vessel's mechanical response. The intima, primarily composed of endothelial cells and a thin layer of extracellular matrix, provides a low-friction interface for blood flow and contributes minimally to the vessel's mechanical stiffness [5]. The media, which contains circumferentially aligned smooth muscle cells and elastin fibres, governs the artery's compliance and ability to recoil under cyclic loading, ensuring physiological adaptation to pulsatile blood pressure [6]. The adventitia, predominantly composed of collagen fibres oriented in a helical arrangement, serves as the primary load-bearing component, providing tensile strength and preventing overexpansion under high-pressure conditions [7]. These structural variations result in a gradient of mechanical properties, where the media facilitates elasticity and the adventitia enhances overall tensile resistance, enabling arteries to withstand haemodynamic forces while maintaining their functional integrity [8].

To replicate the layered architecture and functional biomechanics of native vessels, biomaterials must accommodate both compliance and strength while remaining biocompatible and processable [9]. Hydrogels are promising candidates in this context, offering high water content, tuneable viscoelastic properties, and an ability to mimic the extracellular matrix [10,11]. These attributes make them particularly suitable for vascular graft applications, where dynamic mechanical performance and cellular integration are essential [12].

Among emerging biomaterial inks, polyvinyl alcohol cryogel (PVA-C) stands out for vascular applications due to its combination of printability, biocompatibility, and mechanical tunability [13]. Unlike protein-based or naturally derived hydrogels that often require complex crosslinking chemistries or lack structural integrity [14], PVA-C can be physically crosslinked through freeze–thaw cycles, enabling reproducible solvent-free processing and consistent mechanical performance [13]. Its relatively high viscosity and ability to gel under sub-zero conditions make it well suited for extrusion-based 3D printing, in contrast to lower-viscosity materials such as collagen or hyaluronic acid, which typically require support structures during printing [14,15].

Previous studies have shown that the number of freeze–thaw cycles influences mechanical properties such as stiffness and elasticity [16], and that the addition of bioactive components like gelatine can enhance cell adhesion and hemocompatibility without compromising structural fidelity [17]. PVA-C also exhibits favourable viscoelastic behaviour under cyclic loading, including damping and elastic recovery, positioning it as a strong candidate for replicating the dynamic mechanical environment of native arterial tissues [18].However, achieving directional (anisotropic) mechanical responses remains a challenge, as traditional fabrication methods like casting typically yield isotropic structures that do not reflect the direction-dependent properties of arteries [18].

Bioprinting is an advanced form of additive manufacturing that enables the precise layer-by-layer deposition of living cells, biomaterials, and biologically active molecules to fabricate tissue-like constructs with high spatial precision [15]. It holds considerable promise for regenerative medicine [14], pharmaceutical applications such as drug screening and personalised delivery systems [19], and the development of customised biomimetic implants [15]. By supporting the fabrication of organoids and organ-on-a-chip platforms, bioprinting is also driving innovation in preclinical testing and personalised therapies. Beyond these technical advantages, the approach may significantly improve health-related quality of life by addressing organ shortages, reducing transplant rejection risk, and offering patient-specific therapeutic options [20]. However, the integration of living biological materials raises complex challenges around intellectual property ownership, ethical regulation, and commercialisation, underscoring the need for adapted legal frameworks [15].

These broader developments provide important context for advancing sub-zero bioprinting strategies aimed at engineering vascular grafts with biomimetic mechanical behaviour.

Sub-zero 3D bioprinting, of PVA-C, has emerged as a promising strategy to fabricate layered constructs with controlled internal architectures [21]. Unlike conventional fabrication techniques, layer-by-layer deposition allows for the accurate spatial organisation of material, enabling controlled anisotropic mechanical behaviour that more closely resembles the natural structure of arterial tissues [18]. However, prior research has largely focused on layer thickness and FTCs, with limited studies addressing the effect of multi-layered architectures.

Beyond anisotropic properties, arterial tissues also exhibit viscoelastic and hyperelastic behaviour, meaning they can store, dissipate, and recover mechanical energy under cyclic and large-strain deformation. Viscoelasticity is essential for accommodating dynamic blood flow and reducing mechanical fatigue over time. The storage modulus (E') quantifies the material's ability to store elastic energy and resist deformation under load, directly correlating with structural integrity. The loss modulus (E'') quantifies the material's viscous behaviour, reflecting the material's damping capability and ability to absorb mechanical stress. The ratio of these parameters, expressed as tan δ (E''/E'), indicates the damping efficiency of the material. A higher tan δ suggests greater damping, meaning more energy is lost as heat, whereas a lower value reflects a predominantly elastic response with efficient energy recovery. Materials with frequency-dependent moduli exhibit stronger viscoelastic effects, whereas frequency-independent behaviour suggests a dominant elastic response [14,15,19,22].

Hyperelasticity describes materials that undergo significant deformation but return to their original shape without permanent damage, a characteristic essential for soft biological tissues such as arteries [10]. The stress–strain response of hyperelastic materials is nonlinear, requiring constitutive models beyond simple linear elasticity. In previous studies, PVA-C displayed hyperelastic properties due to its crosslinked polymeric network, making it a promising candidate for biomimetic vascular grafts [23].

The aim of this study is to investigate the influence of increasing the number of layers on the elastic response, and the hyperelastic and viscoelastic behaviour of PVA-C samples under a uniaxial load. By assessing stiffness, frequency-dependent mechanical responses, and nonlinear stress–strain behaviour, this work provides insight into the role of layered architectures in 3D-printed cryogels towards the development of functionally graded PVA-C structures for vascular applications.

2. Materials and Methods

2.1. Overview

The bioprinting of hydrogels is well acknowledged to require a large amount of process parameter optimisation in order to obtain a balance between the printability and structural integrity of the construct [24]. The parameters used in this study, for example, PVA concentration and flow rate, were fixed after a large amount of empirical testing by isolating each material and process parameter and exploring the impact of parametric variation on the process. Further to this, the environmental parameters, which, in this study, were not controlled directly, were monitored to assess their impact.

2.2. Biomaterial Ink Preparation

A solution of 11% *w*/*w* PVA (146–186 kDa, 99+% hydrolysis, Sigma-Aldrich, St. Louis, MO, USA) was dissolved in deionised water by autoclaving for 1 h at 121 °C, followed by mechanical stirring at 50 °C for 1 h, with additional stirring at room temperature (RT) (22.5 ± 1 °C) for 1 h to ensure complete dissolution and homogenisation. The 11%

concentration was selected for its balance between mechanical performance and printability. Autoclaving was specifically incorporated to facilitate dissolution while ensuring that the process can be translated into a sterile manufacturing protocol for future biomedical applications, where sterility is a critical requirement [25].

2.3. Sample Design and Fabrication

Rectangular samples with nominal design dimensions of 12.3 mm × 10 mm and thicknesses of 1, 2, and 3 mm were created using Fusion 360 (Autodesk, San Francisco, CA, USA) and converted to G-code using REGEMAT 3D Designer slicing software (REGEMAT 3D S.L., Granada, Spain). These dimensions were determined based on computer-aided design (CAD) specifications. The 3D input model used for the fabrication process is provided in Appendix A.2 (Figure A1) to illustrate the base CAD geometry. However, to account for potential deviations due to fabrication and processing, the actual dimensions of the samples were measured after the completion of the FTCs and prior to mechanical testing.

A 0–90° infill pattern was implemented to enhance structural stability and ensure uniform mechanical properties across the sample cross-section. Samples were fabricated using a Regemat BIO V1 bioprinter (REGEMAT 3D S.L., Granada, Spain) equipped with a 0.58 mm nozzle. Printing parameters—including flow rate, infill speed, and print bed temperature—were optimised empirically to accommodate the viscosity of the PVA solution. The flow rate was set to 0.6 mm/s, with an infill speed of 2 mm/s. During fabrication, the print bed was maintained at approximately −2.5 ± 1.5 °C, with temperature actively monitored throughout the process. Full details of the monitoring setup and environmental controls are described in Section 2.3, while the recorded temperature profiles during printing are reported in Section 3.6.

Following fabrication, the samples underwent three freeze–thaw cycles (FTCs) alternating between −20 °C and RT (~22 °C), based on protocols established to promote mechanical stability and flexibility [26–28]. Each cycle consisted of freezing at −20 °C for 16 h followed by thawing at RT for 8 h to induce physical crosslinking. Post printing, the samples were immersed in deionised water for four days to achieve full swelling prior to mechanical testing. After swelling, final sample dimensions were recorded; width and thickness were measured directly, whereas the length was not included, as tensile testing was performed based on a fixed gauge length of 10 mm (distance between clamps). Full dimensional data are provided in Appendix A.3.

Samples were produced with 2, 4, and 6 alternating layers, with toolpaths alternating between parallel and perpendicular orientations. Each sample type was fabricated in quadruplicate (n = 4). Table 1 provides a summary of the sample types and corresponding identifiers used in this study.

Table 1. Summary of manufactured sample types.

Sample Type Name	Sample Numbers	Layers	Dimensions (mm^3)
2L-3FTC	2L-1, 2L-2, 2L-3, 2L-4	2	12.3 × 10 × 1
4L-3FTC	4L-1, 4L-2, 4L-3, 4L-4	4	12.3 × 10 × 2
6L-3FTC	6L-1, 6L-2, 6L-3, 6L-4	6	12.3 × 10 × 3

2.4. Dimensional Fidelity Analysis

Prior to the mechanical characterisation, the reproducibility of the printed constructs was assessed; the actual width and thickness of all samples were measured and compared against the CAD specifications for each sample type. The target digital dimensions were 12.3 mm (width) × 1 mm (thickness) for 2L samples, 12.3 mm × 2 mm for 4L samples, and

12.3 mm × 3 mm for 6L samples (Appendix A.2, Figure A1). The average dimensional deviations, calculated from four printed samples per group (Appendix A.3, Table A2), were as follows:

2L samples: width deviation = 1.6 ± 2.8 %, thickness deviation = −23.4 ± 11.5 %;

4L samples: width deviation = −1.4 ± 3.3 %, thickness deviation = −10.5 ± 1.6 %;

6L samples: width deviation = −5.9 ± 3.2 %, thickness deviation = −18.5 ± 3.4 %.

While deviations in thickness were more pronounced—likely due to material spreading and layer compaction—width measurements showed acceptable reproducibility, remaining within ±6% of the target. These findings indicate that, despite the thermal and structural challenges of sub-zero printing, dimensional fidelity in the printing plane was retained across constructs.

2.5. Sample and Environmental Monitoring

The ambient environmental conditions during fabrication were carefully monitored to ensure consistency. The printing bed temperature was tracked using a LabJack T7 thermocouple (Labjack, Lakewood, CO, USA), while ambient temperature and humidity were recorded to account for potential variations in the fabrication environment.

To monitor any visible changes in the structure during fabrication and testing, samples were imaged at three stages: post bioprinting (within 15 min), post FTCs, and during mechanical testing at 40% strain. A Dino-Lite Digital Microscope (Dino-Lite, Torrance, CA, USA) was used to capture high-resolution images, enabling the visual analysis of the structural integrity and deformation patterns of the samples.

Samples were weighed before and after each dynamic mechanical analysis (DMA) test repetition to quantify dehydration effects, as these tests were significantly longer in duration than the ramp tests and, thus, represent a more significant risk of dehydration. The percentage weight change was calculated relative to the initial sample weight to assess hydration loss during testing. Measurements were conducted for the first two DMA test repetitions, and, as weight loss remained consistent between them, further measurements were not performed for the third repetition.

2.6. Mechanical Testing and Data Processing

2.6.1. Uniaxial Tensile Testing

Tensile tests were performed using a Bose ElectroForce 3200 testing system (TA Instruments, New Castle, DE, USA) equipped with a 5 N load cell, with data acquisition managed via WinTest software (TA Instruments, New Castle, DE, USA). Samples were clamped at both ends with a gauge length of 10 mm and subjected to uniaxial ramp testing up to a maximum strain of 40% at a constant displacement rate of 0.25 mm/s. Load–displacement data were recorded and converted to stress–strain curves using standard engineering definitions.

Uniaxial tensile tests were performed on samples with a total length of 30 mm (10 mm gauge length and 10 mm per clamping region) and a width of 12.3 mm. Each sample was subjected to five loading cycles; the first cycle was used for preconditioning and was excluded from subsequent analysis. The average and standard deviation (SD) were calculated from the final four repetitions (T2–T5). Environmental testing conditions were not actively controlled; however, all experiments were conducted under nominal ambient laboratory conditions (approximately 21–23 °C and 40–50% relative humidity).

2.6.2. Stress–Strain Data Processing and Young's Modulus Calculation

Young's modulus (YM) was calculated at predefined strain levels (10%, 20%, 30%, and 40%) using a custom MATLAB R2023a (The MathWorks Inc., Natick, MA, USA) script, which incorporated error estimation for stress and strain values. A point-wise approach was applied within ±0.5% strain of each target level to determine YM. The mean, SD, and propagated error were computed for each test repetition to assess variability and reliability.

Error propagation was employed to quantify measurement uncertainties in the stress–strain data at each recorded point. Standard error propagation techniques were applied, incorporating uncertainties from multiple sources. The force measurement uncertainty of ±0.025 N was derived from the specifications of the load cell, while the displacement uncertainty of ±0.002 mm was based on the mechanical testing machine specifications. Additionally, the cross-sectional area uncertainty was determined from the calliper precision (±0.02 mm).

Linear and quadratic regressions were applied to the stress–strain data to determine the most accurate representation of the material's mechanical response. The linear fit used the equation $\sigma = a\varepsilon + b$, while the quadratic fit used $\sigma = c\varepsilon^2 + d\varepsilon + e$, where σ is the engineering stress, ε is the engineering strain, and a–e are fitting parameters. The best-fitting model was selected based on the highest coefficient of determination (R^2) and minimal residuals. Details of the R^2 values and regression analysis are provided in Appendix B.2.

2.6.3. Dynamic Mechanical Analysis

DMA was performed to assess the viscoelastic behaviour of PVA-C samples by measuring the storage modulus (E′) and loss modulus (E″) across a physiologically relevant frequency range. These parameters characterise the material's ability to store and dissipate mechanical energy. Additionally, the complex modulus (E*) was calculated as follows:

$$E^* = \text{sqrt}\,(E'^2 + E''^2), \tag{1}$$

where E* represents the total mechanical stiffness, combining both the elastic and viscous contributions. The phase angle (δ), which is derived from the ratio of E″ to E′, indicates the degree of viscoelasticity, with larger δ values representing increased damping.

The loss tangent (tan δ), also referred to as the damping factor, was calculated as follows:

$$\tan\delta = E''/E', \tag{2}$$

where tan δ provides insight into the balance between elastic and viscous behaviour. A lower tan δ suggests a more elastic response, whereas a higher tan δ indicates increased energy dissipation.

DMA tests were conducted at RT under controlled hydration conditions. Samples were subjected to sinusoidal displacement at frequencies ranging from 1 to 10 Hz, simulating cyclic deformations experienced by arterial tissues [29]. The storage modulus (E′) represents the elastic component of the response, while the loss modulus (E″) reflects the viscous contribution. The testing parameters, including frequency, mean displacement levels, and dwell times, are summarised in Table A3 in Appendix B.1.

Each sample underwent three repeated tests on different days. Between tests, samples were rehydrated in deionised water for 24 h to maintain consistent swelling behaviour and mechanical properties. Lower frequencies, such as 0.5 Hz, were excluded to minimise dehydration effects. After completing the frequency sweep, samples were returned to the initial frequency of 1 Hz to assess any changes in mechanical response over the course of testing.2.7. Data Analysis

All experimental data were processed using MATLAB and Microsoft Excel (Microsoft, Redmond, WA, USA). Engineering stress and strain, along with dynamic mechanical properties, were computed using the equations summarised in Table 2. These include calculations for stress, strain, dynamic stiffness, storage and loss stiffness, shape factor, and the final derivation of storage (E′) and loss (E″) moduli.

Force and displacement data from uniaxial tensile testing were used to calculate stress and strain using Equations (3)–(5). Stiffness values derived from DMA were processed via Fourier analysis to determine phase lag (δ), from which storage and loss stiffness values were obtained using Equations (8)–(10). These were normalised using the shape factor (Equation (10)) to compute E′ and E″ (Equations (11) and (12)).

Table 2. Summary of mechanical equations used for data analysis.

Equation No.	Formula	Description	Variables
(3)	$\sigma = F/A$	Engineering stress	σ: stress (Pa); F: force (N); A: cross-sectional area (mm^2)
(4)	$A = w \times T$	Cross-sectional area	w: width (mm); T: thickness (mm)
(5)	$\varepsilon = d/L$	Engineering strain	ε: strain; d: displacement (mm); L: gauge length (mm)
(6)	$E = \sigma/\varepsilon$	Young's modulus	E: modulus (Pa); σ: stress (Pa); ε: strain
(7)	$K^* = F/d$	Dynamic stiffness	K*: dynamic stiffness (N/mm); F: force (N); d: displacement (mm)
(8)	$K' = K^* \times \cos(\delta)$	Storage stiffness	K′: storage stiffness (N/mm); δ: phase angle (radians)
(9)	$K'' = K^* \times \sin(\delta)$	Loss stiffness	K″: loss stiffness (N/mm); δ: phase angle (radians)
(10)	$S = (w \times T)/h$	Shape factor	S: shape factor; h: sample height (mm)
(11)	$E' = K'/S$	Storage modulus	E′: storage modulus (Pa)
(12)	$E'' = K''/S$	Loss modulus	E″: loss modulus (Pa)

Statistical analysis was conducted to evaluate intra- and inter-sample variability, with data presented as mean ± SD. A one-way analysis of variance (ANOVA) was performed to determine whether significant differences existed among sample groups. ANOVA assesses whether the means of multiple groups are statistically distinct by analysing the variance within and between groups [30]. The significance of observed differences was evaluated using *p*-values, which quantify the evidence against the null hypothesis; smaller *p*-values indicate stronger evidence that the groups are different [31]. If significant differences were detected ($p < 0.05$), Tukey's post-hoc test was applied to conduct pairwise comparisons. Tukey's test accounts for multiple comparisons, allowing for a more reliable identification of significant differences among sample groups [32].

To assess the effect of layer number on viscoelastic properties, ANOVA and Tukey's post-hoc tests were conducted at three distinct frequencies: 2 Hz, 6 Hz, and 10 Hz. This approach allowed for us to evaluate whether differences in storage modulus (E′) and loss modulus (E″) were frequency-dependent and whether the mechanical response varied significantly between the 2L, 4L, and 6L samples at these key frequencies.

3. Results

3.1. Sample Imaging

A representative example of Sample 4L-3 is shown in Figure 1. Figure 1a shows the sample 15 min post bioprinting, with a uniform grid-like structure and well-defined printed filaments. Figure 1b presents the sample after three FTCs, where an increase in opacity and a slight reduction in pore size can be observed. Figure 1c captures the sample at maximum strain (40%) during tensile testing. These images illustrate the structural changes in the samples throughout the fabrication and mechanical testing processes. Images of all samples at each stage are provided in Appendix A.1.

Despite reaching 40% strain, no visible filament breakage or detachment was observed, suggesting that the material maintained structural continuity under tensile loading. Differences in filament separation and structural uniformity between sample types were evident across all three stages, highlighting the influence of layer number on mechanical integrity.

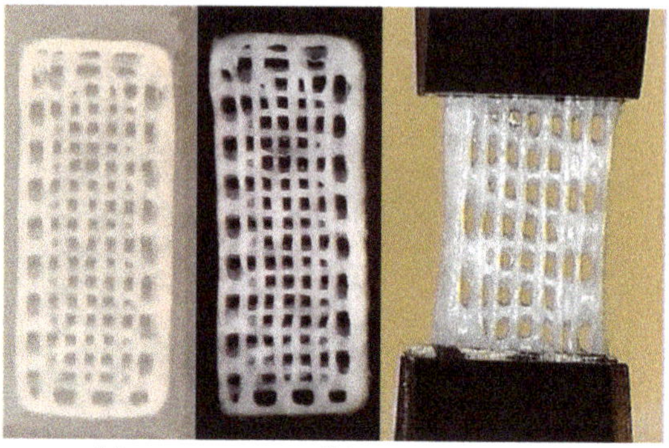

(a) (b) (c)

Figure 1. Representative sample (4L-3) imaged at three key stages of the fabrication and testing process. (**a**) Post bioprinting (15 min after deposition), showing well-defined filament architecture and grid-like infill. (**b**) Post freeze–thaw cycles (FTCs), indicating increased opacity and reduced pore size due to crosslinking. (**c**) During uniaxial tensile testing at 40% strain, revealing uniform deformation without filament detachment. Images captured with a Dino-Lite Digital Microscope under consistent lighting and magnification.

3.2. Stress–Strain Curves and Curve Fitting

Force–displacement data from the ramp tests were converted into stress–strain curves, with YM at a specific strain used for comparisons between different samples. Figure 2 shows a representative uniaxial tension response up to 40% strain for Sample 2L-1. The experimental data include the propagated error and a linear ($\sigma = a\varepsilon + b$) and quadratic ($\sigma = c\varepsilon^2 + d\varepsilon + e$) line of best fit. The quadratic model demonstrates a superior fit based on the coefficient of determination (R^2) value.

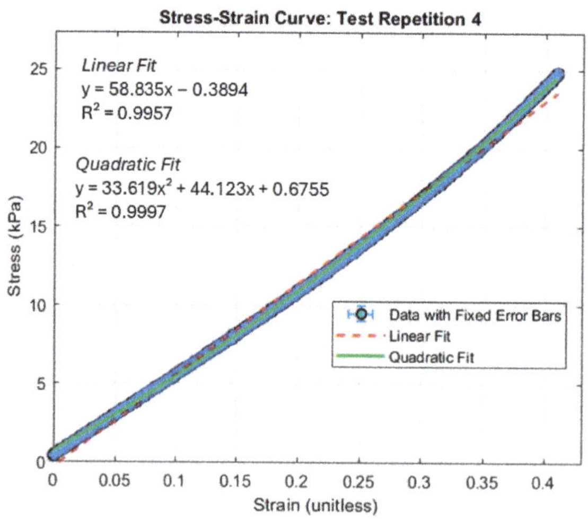

Figure 2. Representative stress–strain response of two-layered PVA-C sample (2L-1) during the fourth uniaxial tensile test repetition. A quadratic fit (R^2 = 0.9997) outperforms the linear model, highlighting slight nonlinear behaviour up to 40% strain. Testing was performed at room temperature, with a displacement rate of 0.25 mm/s, with a 5 N load cell and a 10 mm gauge length. Error bars represent the propagated uncertainty in stress and strain.

The R^2 was used to evaluate the goodness of fit, with values consistently higher for the quadratic model as opposed to the linear model. Table A4 in Appendix B.2 presents the R^2 values for both fitting methods across all samples and test repetitions. All quadratic R^2 values exceeded 0.999, while the values for the linear fits were slightly lower.

3.3. *Young's Moduli*

YM were extracted at 10%, 20%, 30%, and 40% strain. The average YM values, calculated across four test repetitions for each sample type (2L, 4L, and 6L), are shown in Figures 3–5. The values remained consistent across repetitions, as demonstrated by the low SD values, indicating low intra-sample variability. However, significant inter-sample variability was observed across all sample designs.

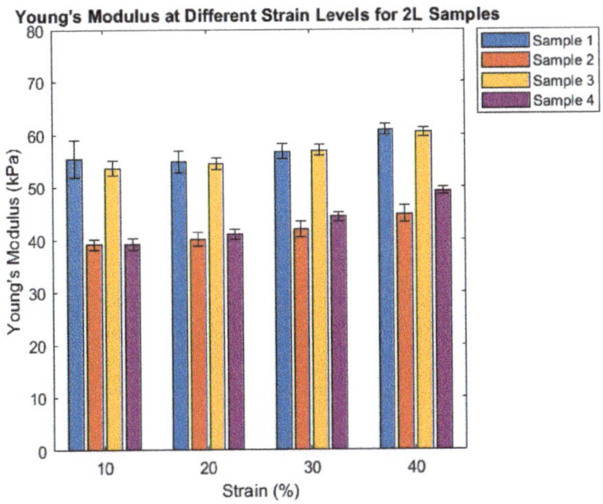

Figure 3. Young's modulus values at predefined strain levels (10%, 20%, 30%, and 40%) for two-layered (2L) PVA-C samples. Each bar represents an individual sample (n = 4 per group), with error bars indicating standard deviation across four repetitions (T2–T5). Measurements were obtained from uniaxial tensile tests using a 5 N load cell at room temperature. YM was calculated using a ±0.5% strain window around each target level.

For each strain level, average YM values and their corresponding SDs for all samples and repetitions are shown in Figure 6, with numerical values provided in Table A5 (Appendix B.3). The analysis reveals an increase in stiffness with strain, with higher modulus values observed at 40% strain compared to 10%, supporting the evidence for a nonlinear (as opposed to linear) mechanical response.

To assess whether layer number significantly influenced YM at different strain levels, a one-way ANOVA was conducted for each strain level (10%, 20%, 30%, and 40%). The analysis revealed statistically significant differences in YM across the sample types (2L, 4L, and 6L) at all strain levels. The full ANOVA results are provided in Tables A7 and A8 (Appendix C.1). Following this, Tukey's post-hoc test was applied to determine which sample types differed significantly. At 40% strain, 2L samples were significantly stiffer than both 4L and 6L samples, while no significant difference was found between 4L and 6L. The complete Tukey comparison results across all strain levels are provided in Tables A9 and A10 (Appendix C).

Within each sample group, some inter-sample consistency was observed. In the 2L group, samples 2L-1 and 2L-3 exhibited similar behaviour, as did samples 2L-2 and 2L-4. In the 4L series, 4L-2 and 4L-3 responded similarly, while, in the 6L series, samples 6L-3 and

6L-4 demonstrated a comparable mechanical response. Although slight variations were observed between individual samples, none were classified as anomalous.

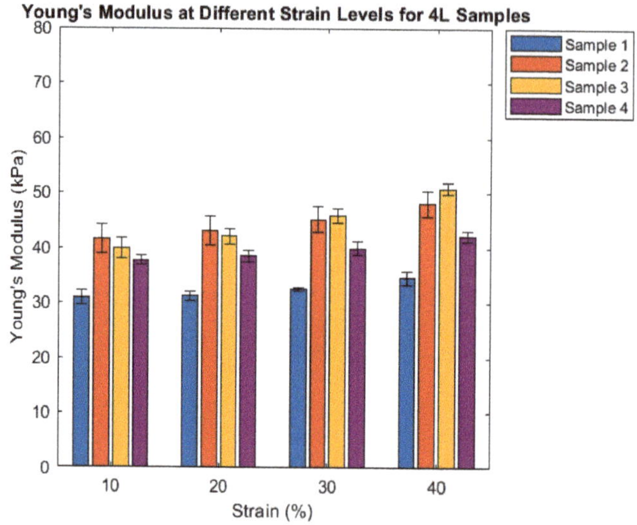

Figure 4. Young's modulus values at predefined strain levels (10%, 20%, 30%, and 40%) for four-layered (4L) PVA-C samples. Each bar represents an individual sample (n = 4 per group), with error bars indicating standard deviation across four repetitions (T2–T5). Measurements were obtained from uniaxial tensile tests using a 5 N load cell at room temperature. YM was calculated using a ±0.5% strain window around each target level.

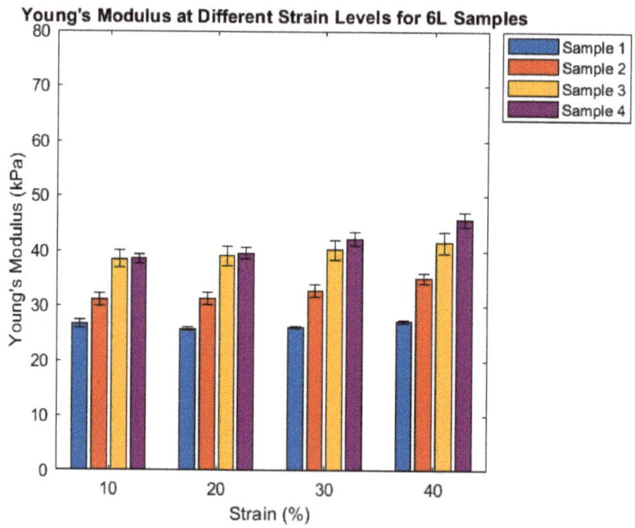

Figure 5. Young's modulus values at predefined strain levels (10%, 20%, 30%, and 40%) for six-layered (6L) PVA-C samples. Each bar represents an individual sample (n = 4 per group), with error bars indicating standard deviation across four repetitions (T2–T5). Measurements were obtained from uniaxial tensile tests using a 5 N load cell at room temperature. YM was calculated using a ±0.5% strain window around each target level.

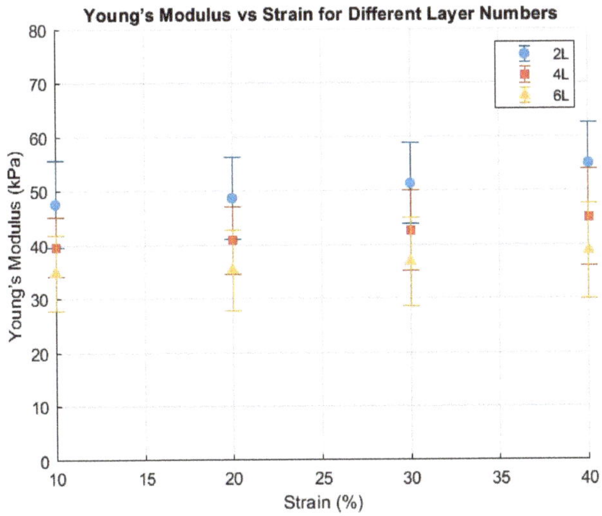

Figure 6. Mean Young's modulus values for 2L, 4L, and 6L PVA-C samples as a function of strain level (10–40%). Data represent the average YM across all four samples per group, with error bars indicating inter-sample standard deviation. All samples were tested under identical conditions (room temperature, 0.25 mm/s displacement rate), and values were extracted from repetitions T2–T5 following preconditioning.

3.4. Dynamic Mechanical Analysis Results

The average storage moduli (E′) and loss moduli (E″) within the range of 1–10 Hz for all of the 2L, 4L, and 6L samples are shown in Figures 7 and 8. The storage and loss moduli for each sample, for all repetitions, are shown in Appendix B.4, Table A6. Figures 7 and 8 show the trends of a decrease in storage moduli and increase in loss moduli with increasing frequency. However, the SDs of the data are large enough to suggest that these trends are not significant. Looking at each sample in isolation, in Appendix B.4, it can be shown that there are large differences in the average moduli between samples, creating very large SD in the combined data sets overall.

Comparing Figures 7 and 8, the storage modulus is consistently higher than the loss modulus across all sample designs and frequencies. At 2 Hz, the E′ increased with the number of layers: the two-layered (2L) samples exhibited the lowest (~8422 Pa), followed by the four-layered (4L) samples (~11,343 Pa) and the six-layered (6L) samples (~10,813 Pa). Similar trends were observed at 6 Hz and 10 Hz, though the differences were less pronounced at higher frequencies.

Statistical analysis confirmed these observations. One-way ANOVA at 2 Hz, 6 Hz, and 10 Hz indicated significant differences between sample types for E′ (Appendix C.2, Table A11). Post-hoc Tukey's Honestly Significant Difference (HSD) tests showed that the 2L samples were significantly different from both 4L and 6L samples at all frequencies, while the difference between 4L and 6L was not significant (Appendix C.2, Table A10). This suggests that increasing the number of layers from two to four significantly increases the modulus, but adding more layers beyond four does not lead to a further significant increase in E′.

For the loss modulus, ANOVA results showed significant differences at 2 Hz and 10 Hz (Appendix B, Table A12), but not at 6 Hz. Tukey's HSD test revealed that, at 2 Hz and 10 Hz, the 2L samples were significantly different from both the 4L and 6L samples ($p < 0.05$), whereas the difference between 4L and 6L remained statistically insignificant.

This suggests that layering affects energy dissipation at lower and higher frequencies but has a diminished effect at intermediate frequencies. At the end of each test, when the frequency was returned to 1 Hz, the storage and loss modulus values remained within the initial measurement range, with differences falling within the SD.

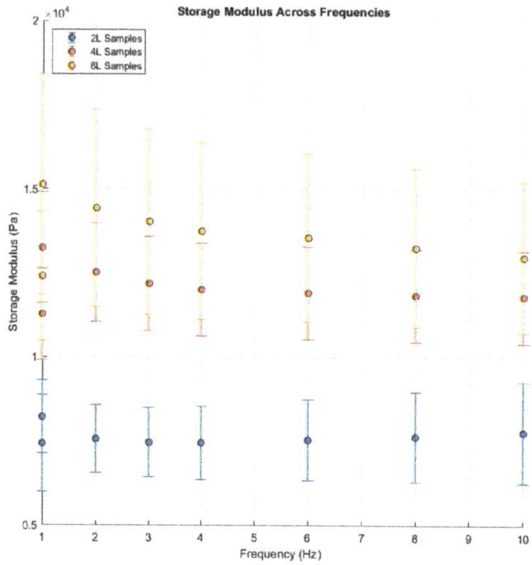

Figure 7. Storage modulus (E′) of all sample types as a function of frequency (1–10 Hz). Data points represent storage modulus (E′) of layered PVA-C constructs (2L, 4L, 6L). Each data point represents the average E′ across four independently fabricated samples, with error bars showing standard deviation. DMA tests were performed at room temperature, with sinusoidal oscillations across 1–10 Hz.

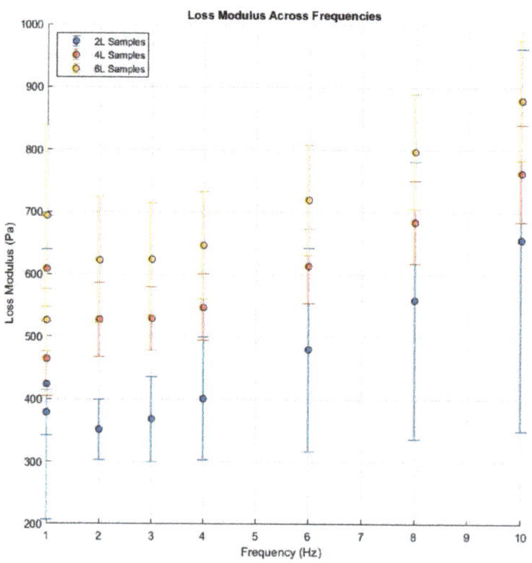

Figure 8. Loss modulus (E″) of all sample types as a function of frequency (1–10 Hz). Data points represent the loss modulus (E″) of layered PVA-C constructs (2L, 4L, 6L). Each data point reflects the average E″ across four independently fabricated samples, with error bars indicating standard deviation. DMA tests were conducted at room temperature using sinusoidal oscillations across the 1–10 Hz frequency range.

3.5. Dehydration Analysis

The percentage weight change was calculated to quantify dehydration, with observed losses ranging from −8.21% to −21.63% (outlined in full in Table A19, Appendix D.1). The average weight change varied across sample types and between test days. Pre-test weight measurements on the second testing day were consistent with those recorded before the first DMA test, indicating that samples had sufficient time to reabsorb water and return to their swollen state between test repetitions. Table 3 summarises the mean percentage dehydration and SD for each sample configuration on Day 1 and Day 2 of testing.

Table 3. Mean and standard deviation of dehydration percentage for each sample type across test days.

Sample Type	Mean % Dehydration (Day 1)	SD % Dehydration (Day 1)	Mean % Dehydration (Day 2)	SD % Dehydration (Day 2)
2L	−19.05	1.79	−17.74	3.14
4L	−10.57	1.64	−11.89	1.96
6L	−9.74	1.67	−9.14	0.68

The results highlight that 2L samples exhibited the highest dehydration levels, while 6L samples retained the most water, suggesting that increased layering may enable water retention.

3.6. Environmental Monitoring

The printing bed temperature, laboratory ambient temperature, and humidity were recorded before and during each print to assess environmental stability throughout the fabrication process. The laboratory ambient temperature remained stable at 23 ± 1 °C, while humidity was maintained at 40 ± 1%, indicating minimal fluctuations in environmental conditions.

The printing bed temperature was monitored at a time step of 0.5 s, capturing temperature variations throughout the printing process. The recorded environmental conditions for each sample configuration were as follows:

- Two-layered (2L) samples: Printing bed temperature: −1.84 ± 1.49 °C;
- Four-layered (4L) samples: Printing bed temperature: −2.95 ± 1.24 °C;
- Six-layered (6L) samples: Printing bed temperature: −2.81 ± 0.92 °C.

4. Discussion

4.1. Elastic Response

The YM values calculated at 10%, 20%, 30%, and 40% strain showed a decreasing trend with increasing layer number (Figure 6). The 2L samples exhibited the highest YM values across all strain levels, averaging 45.7 ± 3.2 kPa at 40% strain, while 4L and 6L samples measured 38.2 ± 2.9 kPa and 34.1 ± 2.4 kPa, respectively.

Statistical analysis using one-way ANOVA confirmed that YM values differed significantly between sample types at all strain levels ($p < 0.05$) for the 2L and 4L samples. Tukey's post-hoc test further revealed that 2L samples had significantly higher YM values than both 4L (Table A8, Appendix C.1) and 6L samples at 40% strain, while no significant difference was found between 4L and 6L. These results indicate that layering impacts stiffness, but beyond four layers, additional layering does not further reduce YM in a statistically meaningful way.

The low intra-sample variability (i.e., across repeated tests on the same sample) demonstrates the reproducibility of the characterisation methodology, with the material response being predominantly elastic after the preconditioning cycle. However, high inter-sample variability was observed, particularly in the 2L and 4L samples, despite the extensive empirical testing used to optimise the fabrication process. This suggests that fabrication-related factors, such as variations in printing bed temperature, polymer deposition accuracy, and localised hydration differences, may contribute to mechanical performance variability between samples.

Visual inspection revealed that 2L samples exhibited the least uniform structure, with non-homogeneous polymer distribution and localised drying effects (Table A1, Appendix A.1). During tensile testing, one sample exhibited visible filament breakage (Sample 2L-4), suggesting weak points which could be associated with dehydration and polymer redistribution. In contrast, 6L samples exhibited minimal filament separation, better hydration retention, and greater structural integrity under an applied load (Table A1, Appendix A). These imaging results confirm that layer number influences both uniformity and mechanical performance, with thinner constructs (2L) exhibiting greater structural inconsistencies, while thicker samples (6L) demonstrate improved stability under mechanical loading.

Given the statistically significant reduction in YM with increasing layer number, it is likely that mechanical decoupling at layer interfaces and polymer crosslinking variations contribute to the observed stiffness reduction. This could be due to the increase in the distance from the print bed and, therefore, the increase in the printing temperature as the number of layers increases (discussed further in Section 4.6). However, the lack of a significant difference between 4L and 6L suggests a saturation effect, where, beyond a certain layering threshold, additional layers do not further weaken the construct.

These findings suggest that increased layering mitigates dehydration effects and improves structural homogeneity but reduces overall stiffness. Further investigations should explore whether modifications in fabrication conditions, such as refining deposition accuracy or implementing controlled printing environments, could enhance consistency in mechanical properties and polymer crosslinking.

4.2. Dynamic Response

The DMA results indicate that layering influences the storage moduli (E′) and loss moduli (E″), but that the samples do not exhibit significant frequency dependence. While subtle trends were observed across frequencies, statistical analysis confirmed that the differences fell within the SD, reinforcing the frequency-independent nature of these samples.

Layering had a clear effect on mechanical properties, with 2L samples exhibiting significantly lower E′ values compared to 4L and 6L samples at 2 Hz and 10 Hz ($p < 0.05$, Appendix C.2, Tables A13–A15). However, increasing the number of layers beyond four did not result in a statistically significant increase in stiffness. This may suggest that additional layering reaches a saturation point where further gains in mechanical reinforcement are minimal. A similar pattern was observed for E″, where 2L samples showed significantly different values from 4L and 6L at 2 Hz and 10 Hz, but not at 6 Hz ($p < 0.05$, Appendix C.2, Tables A16–A18). This suggests that layering affects energy dissipation at lower and higher frequencies, but the effect diminishes at intermediate frequencies.

Despite minor fluctuations, Figures 7 and 8 further illustrate this trend, showing that loss modulus (E″) remains substantially lower than storage modulus (E′) across all configurations, confirming their predominantly elastic response. This predominance of elastic behaviour is consistent with the rheological profile expected from shear-thinning materials such as PVA-based hydrogels, where the flow behaviour index is typically <1, reflecting non-Newtonian behaviour under shear [33].

Whilst a trend between frequency and the storage and loss moduli was observed, it was not statistically significant due to the high level of variability, particularly for inter-sample differences. Similarly, the results presented for all samples in Appendix B.4 did not show a distinct trend beyond the inherent variability between sample repeats. This contrasts with native arterial tissues, which exhibit pronounced frequency-dependent viscoelastic properties due to their complex hierarchical structure and active cellular components [29]. The absence of significant frequency dependence in the bioprinted PVA-C samples suggests that, while they provide a stable mechanical response under oscillatory loading, they do not fully replicate the dynamic behaviour of arterial tissues, which progressively stiffen at higher frequencies to accommodate pulsatile blood flow.

While individual 2L, 4L, and 6L samples exhibited frequency-independent behaviour, their differing absolute responses suggest the potential for functionally graded materials (FGMs). By designing alternative spatial distributions of the PVA-C toolpath, normalised for the continuity of thickness, for example, embedded in another material, or dual 3D printing, it may be possible to engineer constructs that enable a change in storage and loss modulus through space. This could enable the tuning of mechanical gradients to enhance biomimetic performance and optimise vascular graft designs for improved physiological compatibility.

4.3. Nonlinear Response

To assess any nonlinearity in the mechanical response, linear and quadratic data fits were undertaken. As shown in Figure 2, while the linear model provided a reasonable fit across most of the strain range, it deviated at both lower and higher strain levels, falling outside the error bars. In contrast, the quadratic model offered a better representation of the stress–strain response (Appendix B.2, Table A4). The improvement in R^2 was minimal, however, and so a hyperelastic framework such as Mooney–Rivlin or Ogden is not required within the tested strain rate (up to 40%) [34]. The hyperelastic effect could become more pronounced beyond the 40% strain range utilised in this study. Future work should explore whether these cryogels exhibit nonlinear responses at higher deformations, where hyperelastic models could become more appropriate [35].

4.4. Comparison with the Literature on PVA-C for Vascular Applications

To contextualise the findings, the mechanical properties of the layered PVA-C samples were compared with the existing literature. Table 4 summarises key values from this study alongside previous work.

The bioprinted PVA-C constructs in this study exhibited YM (30–50 kPa), which are comparable to previous bioprinted samples from Crolla et al. [36] and Gale et al. [37]. However, both casted PVA-C hydrogels and native arterial tissues display considerably higher stiffness. Native arteries exhibit YM values in the range of 90–184 kPa (Dutta et al. [38]) and an ultimate tensile strength up to 1430 kPa (Holzapfel et al. [39]). The range of moduli reported in the literature for vascular tissue is very broad. None of the PVA-C samples replicated the mechanical strength of native tissues, regardless of fabrication method. This suggests that additional reinforcement strategies such as fibre incorporation, composite structuring, or crosslinking strategies may be required to improve mechanical performance for vascular applications.

We also benchmark against widely used synthetic grafts such as expanded polytetrafluoroethylene (ePTFE) and Dacron. These materials are substantially stiffer than both native arteries and the printed PVA-C constructs reported in this study [40]. While their high mechanical strength ensures structural durability, their limited compliance has been associated with suboptimal integration and elevated failure rates, particularly in small-diameter vascular applications [41]. In contrast, the softer mechanical profile of cryogel-based constructs may offer better compliance matching and potentially more favourable haemodynamic performance, especially when enhanced through reinforcement or biofunctionalisation [42].

Table 4. Comparison of Young's modulus and viscoelastic properties with the literature, including native arteries, synthetic grafts, and other PVA-C constructs.

Study	Sample Type	Young's Modulus (kPa)	Storage Modulus (E′) (kPa)	Loss Modulus (E″) (kPa)
Panieraki et al. (current study)	Bioprinted PVA cryogel	30–50	8.02–12.96	0.52–0.57
Crolla et al., 2021 [36]	Bioprinted PVA cryogel (parallel and perpendicular printing orientations)	15.10–58.90	59.20–96.80	1.89–10.2
Crolla et al., 2021 [36]	Casted PVA cryogel (10–15% *w*/*v*)	78–136	60.00–102.00	5.10–10.20
Gale et al., 2024 [37]	Bioprinted PVA cryogel	48.45–73.81	7.44–11.56	0.21–0.71
Burton et al., 2017 [29]	Porcine coronary artery (LAD)	N/A	14,470–25,560	1570–2710
Dutta et al., 2013 [38]	Baboon carotid artery (arterial fibroblasts)	90–184.2	N/A	N/A
Holzapfel et al., 2005 [39]	Human coronary artery (ultimate tensile stress)	Adventitia: 1300–1430 Media: 419–446 Intima: 391–394	N/A	N/A
Lim et al., 2021 [40]	Dacron (synthetic vascular graft)	3,000,000	N/A	N/A
Lim et al., 2021 [40]	ePTFE (synthetic vascular graft)	500,000	N/A	N/A

Unlike the study by Crolla et al. [36], which primarily investigated single-layered and orthogonally stacked configurations, this work specifically examines the effect of progressive layering on mechanical properties, introducing a structured multi-layered design to assess its impact on both elastic and viscoelastic behaviour. This approach enabled a systematic investigation into how incremental layering influences mechanical stability.

The storage moduli (E′) values for the bioprinted samples (10–20 kPa) were lower than those reported in Crolla et al. [36] but comparable to Gale et al. [37]. Casted PVA-C hydrogels displayed significantly higher E′ values (60–100 kPa) [36], suggesting increased crosslinking uniformity and structural rigidity enabled by the casting process. This research extends previous work by assessing inter- and intra-sample variability through statistical analysis, providing a quantitative framework for evaluating sample consistency with respect to fabrication and mechanical characterisation, an aspect that was not extensively explored in previous bioprinting studies. The loss moduli (E″) of the bioprinted PVA-C constructs (5.2–5.7 kPa) were significantly lower than those of native arteries (~15.7–27.1 kPa) [29]. While casted PVA samples exhibited higher E″ values, they also did not reach the range required to mimic the viscoelastic response of biological tissues. However, casted samples remain isotropic, lacking the orthotropy achievable through bioprinting, which is critical for replicating the fibre-reinforced architecture of arterial tissues.

Layered PVA-C structures present a promising approach for fabricating functionally graded biomaterials, which could enable spatially variable mechanical properties that better replicate the behaviour of native vascular tissue. Previous studies have demonstrated that anisotropic fibre alignment plays a crucial role in arterial mechanics [43]. While the current study did not explore the directional alignment of the toolpath, previous studies have investigated this effect [36].

4.5. Study Repeatability and Environmental Conditions

Dehydration levels varied across sample types, with 2L samples exhibiting the highest weight loss (−19.04% on Day 1, −17.74% on Day 2) and 6L samples retaining the most moisture (−9.74% on Day 1, −9.14% on Day 2). These findings aligned with structural imaging observations (Appendix A.1, Table A1), where thinner constructs, particularly Sample 2L-4 at 40% strain, displayed irregular polymer distribution, increased filament separation, and the formation of a localised hole under tensile loading (Sample 2L-4, Table A1, Appendix A.1). While such features suggest that dehydration may affect mechanical integrity, the current data does not establish a direct causal relationship. In contrast, 6L samples maintained greater filament cohesion and overall uniformity, indicating that increased layering supports moisture retention and structural stability. The 4L samples demonstrated intermediate behaviour, with moderate dehydration and partial filament separation, suggesting that four layers may not fully mitigate dehydration-induced structural disruption under strain.

Despite variations in hydration retention, the mechanical variability observed across sample designs suggests that additional factors—such as polymer crosslinking density, deposition uniformity, and filament cohesion—may have influenced performance. For instance, uneven temperature distribution during freeze–thaw cycling can induce local variations in crosslinking density, directly impacting stiffness and elasticity [33]. Irregular deposition may also lead to inconsistent filament packing and weaker interlayer bonding, reducing load transfer efficiency [44]. Furthermore, reduced filament cohesion at the interfaces can compromise structural integrity under mechanical load [45]. These aspects may interact with dehydration effects, contributing collectively to the observed mechanical variability.

While the printing bed temperature was monitored throughout the fabrication process, it was not actively regulated by the bioprinter and fluctuated between −1.84 °C and −2.95 °C. Section 3.6 presents the full temperature profile and the smallest temperature variation occurred in 6L samples. In comparison, Crolla et al. [46] reported an initial bed temperature of −35 °C, followed by a rapid increase. This research has improved on temperature consistency by maintaining a relatively stable sub-zero profile. Nonetheless, the lack of more precise temperature control may have influenced polymer deposition and crosslinking consistency.

Environmental temperature and humidity were also not actively controlled, likely exacerbating dehydration in thinner constructs with higher surface-area-to-volume ratios. While the bed temperature would directly influence the first layer, subsequent layers were exposed to varying thermal conditions dictated by the underlying layer and the surrounding environment. These ambient fluctuations may have led to different vertical gradients in temperature and, therefore, layer-dependent temperature profiles. Such variability could have led to uneven crosslinking and mechanical inconsistencies.

These findings highlight the importance of maintaining hydration and ensuring both in-plane and the perpendicular-axis temperature and print uniformity to preserve structural heterogeneity. Future work should consider using a temperature-controlled chamber to minimise thermal fluctuations and stabilise polymer deposition. Additionally, regulating environmental humidity during both fabrication and storage may further reduce dehydration-induced variability. Formulation strategies—such as incorporating hydration-retentive additives—may also help to improve homogeneity and mechanical robustness, especially in thinner constructs.

Underlying the mechanical variability is the polymer microstructure, which is highly sensitive to environmental conditions during fabrication. Paxton et al. [47] showed that PVA-based inks transition from Newtonian to shear-thinning non-Newtonian behaviour under sub-zero temperatures. This rheological shift affects extrusion fidelity through changes in the consistency index (K) and flow behaviour index, both of which are temperature-dependent [37,38]. Local shear conditions, influenced by nozzle geometry and pressure differentials, further modulate filament deposition and potentially alter polymer chain alignment or local crosslinking density [47].

Freeze–thaw cycling promotes partial crystallisation and intermolecular hydrogen bonding, forming physical crosslinks that reinforce the hydrogel network [37,48]. These processes are affected by parameters such as polymer concentration, freeze–thaw duration, and cooling rate [49] Consequently, fluctuations in ambient or bed temperature (relative to samples and layers) during printing may result in heterogeneous microstructures and crosslinking profiles, contributing to inter-sample variability.

Finally, polymer–solvent interactions strongly govern the hydration behaviour and mechanical performance of PVA cryogels. The hydrophilic nature of PVA enables extensive hydrogen bonding with water, facilitating water uptake and retention [50]. These interactions support swelling and elasticity, while cryogenic processing separates water into amorphous domains, promoting both flexibility and structural integrity [51]. For biomedical applications, this retained water supports nutrient diffusion and mechanical compliance [52–54]. Modulating solvent composition allows for tuning pore size and hydration properties by altering polymer chain mobility [50,55]. Optimising these interactions in future research will be vital for tailoring mechanical behaviour and stability in 3D-printed cryogels.

4.6. Rheological Considerations Under Sub-Zero Printing Conditions

Although rheological characterisation is a critical step in evaluating the printability of biomaterials, it was not feasible in the present study due to the sub-zero printing protocol and the thermoresponsive nature of PVA. The formulation used here undergoes physical crosslinking via freeze–thaw-induced crystallisation, with gelation initiating near the freezing point of water (~0 °C). As a result, the ink transitions rapidly from a viscous fluid to a semi-solid gel within a narrow temperature window, making conventional rheological measurements using cone-plate rheometers impractical at printing-relevant sub-zero conditions. Performing controlled shear rate sweeps or oscillatory tests below 0 °C is technically challenging due to rapid gelation and thermal instability.

Nonetheless, previous studies have characterised the rheological behaviour of comparable PVA-based systems. PVA solutions are known to exhibit temperature-sensitive shear-thinning behaviour, where the flow behaviour index decreases as temperature approaches freezing, indicative of a transition from Newtonian to non-Newtonian flow [47,56]. Gelatin–PVA blends have shown similar trends, with viscosity decreasing under applied shear but recovering post deposition, facilitating stable filament formation during extrusion [57]. In the context of sub-zero bioprinting, Crolla et al. [46] demonstrated that directional solidification during extrusion induces orthotropic mechanical behaviour in PVA cryogels, attributed to the formation of filament-level mesostructures. This study revealed that samples printed with smaller nozzle diameters exhibited higher stiffness along the print direction and increased orthotropy, which were linked to greater filament alignment and reduced inter-filament bonding. While direct rheological measurements were not performed, these findings highlight the significant influence of sub-zero extrusion parameters on the microstructural organisation and resultant mechanical performance of cryogels.

Although rheological data were not directly measured in this study, the qualitative printing performance—minimal filament spreading, consistent extrusion, and acceptable dimensional fidelity—suggests that the ink behaved in line with expected shear-thinning and recovery characteristics reported in the literature. The overall mean dimensional deviation of 11.4 ± 3.6% supports the reliability of the sub-zero extrusion-based manufacturing process.

4.7. Study Reflections

This study provides valuable insights into the influence of layering on the mechanical properties of 3D-printed PVA-C; however, several limitations should be considered. The absence of anisotropic fibre reinforcement remains a key constraint, as fibre orientation plays a critical role in replicating the mechanical behaviour of arterial tissues [58].

While anisotropy is an important characteristic of native arteries, the mechanical tests performed in this study were uniaxial and conducted in a single direction. As such, the results reflect only the bulk mechanical response along that axis and do not capture potential directional dependencies in stiffness or energy dissipation. A single YM is not a sufficient descriptor for an anisotropic system; biaxial, shear and/or multi-directional tensile testing, are required to fully characterise the anisotropic mechanical behaviour of layered PVA-C constructs.

Additionally, while measurements taken during the DMA captured short term dehydration effects, long-term hydration characteristics under mechanical testing and physiological conditions remain unverified. Although weight measurements confirmed consistent reswelling between test repetitions, further studies should investigate hydration-dependent mechanical performance, particularly in in vitro conditions, to determine suitability for vascular graft applications.

Mechanical characterisation was limited to a maximum strain of 40%. To fully capture the nonlinear mechanical response of layered PVA-C, future work should include higher strain testing. Furthermore, testing should explore the yield point to evaluate ultimate tensile strength and failure behaviour. Extended cyclic loading experiments are needed to assess fatigue resistance and long-term durability under physiologically relevant conditions.

A more detailed evaluation of geometry retention and interlayer cohesion at the filament level—such as filament width uniformity, pore closure, or potential delamination—would require high-resolution imaging techniques like SEM. Such analyses were beyond the scope of this mechanical-focused investigation, but represent an important future direction for optimising the structural stability of multi-layered cryogel systems.

Further exploration of FGM designs is warranted. While individual 2L, 4L, and 6L samples exhibited frequency-independent behaviour, exploring the combination of these configurations could generate a structure with dynamic moduli which are spatially variable. Such an approach could enhance biomimetic performance by leveraging the distinct dynamic stiffness contributions of each layered configuration.

To translate the findings of this study into cylindrical geometries representative of arterial grafts, alternative printing strategies would need to be considered. Printing along the axial (Z) direction would, in theory, allow for vertical layering; however, under sub-zero printing conditions, this configuration compromises bed temperature control, which is essential for maintaining cryogel integrity. A more viable approach involves deposition onto a rotating cylindrical substrate, which may enable radial layering while preserving thermal stability [59,60]. Dell et al. [59] demonstrated a mandrel-based bioprinting system capable of fabricating tubular hydrogel constructs with controlled fibre orientation and multi-layered architecture, while Reeser et al. [60] reviewed additive-lathe 3D printing techniques that facilitated the construction of cylindrical structures with improved mechanical fidelity and reduced reliance on support materials. These approaches underscore the potential of radial layering to modulate mechanical properties along circumferential and axial axes, aligning with the functionally graded design principles investigated in this study. Future work should explore the adaptation of this strategy to bioprinted PVA-C, with the aim of more accurately replicating the anisotropic mechanical environment of native arteries.

Although biological validation was beyond the scope of this study, the biocompatibility and integration potential of polyvinyl alcohol cryogel (PVA-C) constructs are well supported in the literature. Due to the sub-zero bed temperature and the repeated freeze–thaw cycles required for physical crosslinking, the current process is incompatible with cell encapsulation during printing, as these conditions would compromise cell viability. However, post-fabrication strategies have been shown to enhance cellular compatibility. For instance, improved endothelial cell adhesion and spreading have been observed in PVA–gelatine composites [61], and blending PVA with gelatine—providing cell-adhesive arginine–glycine–aspartic acid sequences—has been shown to support endothelialisation [27]. Additional approaches, such as surface functionalisation with adhesive peptides and perfusion-based seeding, may further promote biological integration after fabrication [62]. The influence of freeze–thaw cycling on cytocompatibility has also been explored, with reports indicating that multiple cycles increase porosity and mechanical strength without impairing cell viability [63]. Collectively, these findings support the feasibility of post-printing cell seeding strategies for vascular tissue engineering using PVA-C scaffolds.

5. Conclusions

This study explored the mechanical properties of sub-zero-bioprinted multi-layered PVA-C as candidate materials for vascular graft applications. The results demonstrated that increasing the number of layers led to a moderate reduction in stiffness, likely due to decoupling at layer interfaces, and local variations in crosslinking induced by thermal gradients during fabrication. All constructs exhibited a predominantly elastic response under cyclic loading, with minimal frequency dependence of the storage modulus (E').

The testing methodology proved robust and reproducible across repetitions, although inter-sample variability was observed. This variability underscores the influence of fabrication conditions—such as environmental fluctuations, hydration retention, and print uniformity—on mechanical performance and highlights the importance of further optimisation in the bioprinting process.

Statistical analysis confirmed that both storage and loss moduli differed significantly between two-layered and multi-layered constructs at 2 Hz and 10 Hz, indicating that layering influences energy dissipation at specific frequency ranges. However, beyond four layers, additional layering did not significantly alter mechanical performance, suggesting a saturation effect.

Compared to the prior literature, the layered bioprinted PVA-C constructs achieved similar mechanical performance, though they remain less stiff than native arterial tissues. Importantly, bioprinting offers the advantage of anisotropic control, an essential feature not achievable with casted hydrogels. This study was limited to uniaxial testing, capturing only the bulk response in a single direction. Since arterial tissues are inherently anisotropic, future investigations should include biaxial or shear testing as well as in vitro experiments under physiological flow and pressure conditions to better reflect dynamic vascular environments.

Additionally, incorporating cellular components and biofunctionalisation strategies—such as endothelialisation or growth factor delivery—could improve the biological performance of these constructs. Translational efforts should also consider the long-term stability, sterilisation compatibility, and scalability of the fabrication method to support clinical application.

Finally, the distinct mechanical responses of the 2L, 4L, and 6L samples suggest that spatially arranging different layered configurations could be a promising strategy for creating functionally graded materials (FGMs) with tuneable mechanical properties. Such designs may enable more biomimetic vascular grafts by better replicating the heterogeneous structure and variable mechanical response of native arteries.

Author Contributions: A.P.: Writing—original draft, Visualisation, Validation, Project administration, Methodology, Investigation, Formal analysis, Data curation. N.M.: Writing—review and editing, Supervision, Methodology, Data curation. C.A.: Writing—review and editing, Supervision, Methodology, Data curation. R.J.D.: Writing—review and editing, Supervision, Methodology, Data curation. L.E.J.T.-S.: Writing—review and editing, Validation, Supervision, Resources, Methodology, Funding acquisition, Formal analysis, Conceptualisation. All authors have read and agreed to the published version of the manuscript.

Funding: This work was supported by the UK Engineering and Physical Sciences Research Council (EPSRC) grant number EP/S036717/1.

Data Availability Statement: The data presented in this study are available on request from the corresponding author. Currently, the data are not publicly archived due to practical constraints. No ethical or privacy restrictions apply.

Acknowledgments: The authors sincerely thank Marc Granado-Gonzalez for his guidance in refining the statistical calculations and methodological framework. His insights into error propagation, curve fitting, and Young's modulus estimation contributed to a more rigorous and scientifically robust analysis, enhancing the accuracy and reliability of the mechanical property assessments in this study.

Conflicts of Interest: The authors declare no conflicts of interest.

Abbreviations

The following abbreviations are used in this manuscript:

PVA-C	Polyvinyl alcohol cryogel
3D	Three-dimensional
FTCs	Freeze–thaw cycles
RT	Room temperature
DMA	Dynamic mechanical analysis
YM	Young's modulus
E'	Storage modulus
E''	Loss modulus
E^*	Complex modulus
$\tan\delta$	Loss tangent
FGM	Functionally graded material
ANOVA	Analysis of variance
SD	Standard deviation
HSD	Honestly significant difference
R^2	Coefficient of determination
ε	Engineering strain
σ	Engineering stress
K'	Storage stiffness
K''	Loss stiffness
K^*	Complex (dynamic) stiffness

δ	Phase angle
S	Shape factor
A	Cross-sectional area
F	Force
d	Displacement
L	Gauge length
w	Sample width
T	Sample thickness
h	Sample height
2L	Two layered
4L	Four layered
6L	Six layered

Appendix A. Sample Imaging and Fabrication Data

Appendix A.1. Sample Imaging at Key Stages

Figures and tables corresponding to the following structural changes in samples:

- Post bioprinting;
- Post freeze–thaw cycles (FTCs);
- During mechanical testing (40% strain).

Table A1. Samples imaging post printing, after FTCs, at 40% (maximum) strain.

2L-1			2L-2		
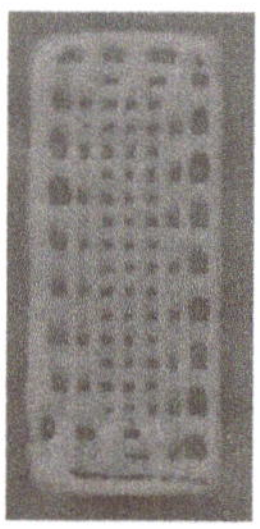	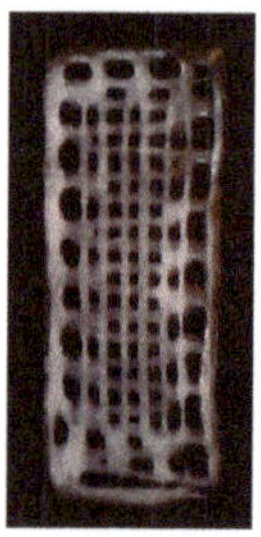	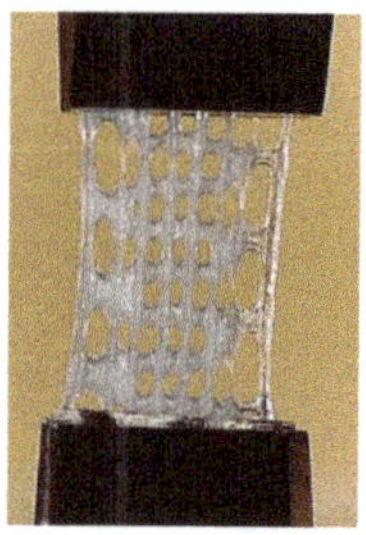	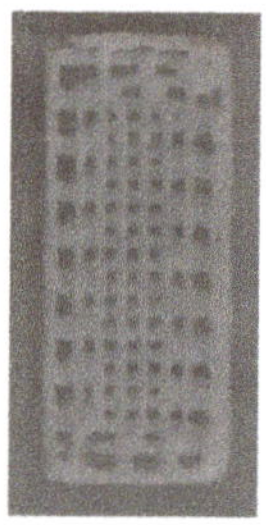	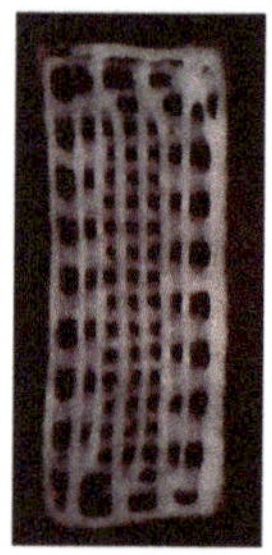	
2L-3			2L-4		
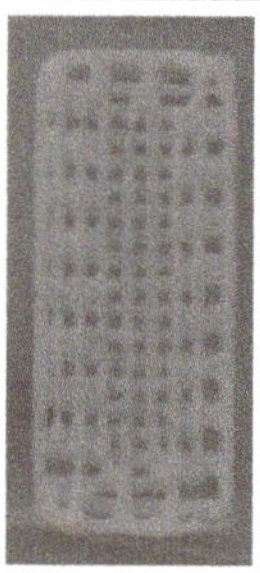		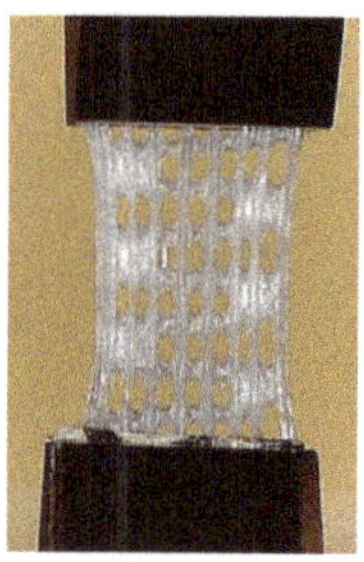		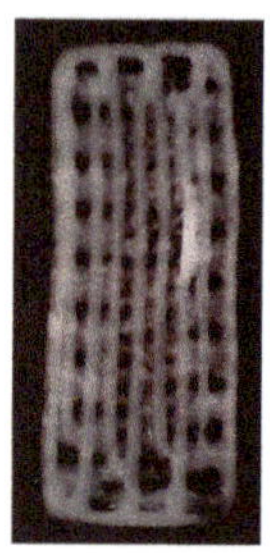	

Table A1. *Cont.*

4L-1	4L-2
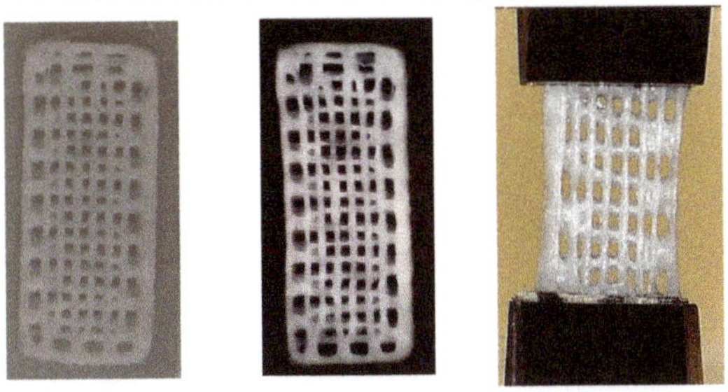	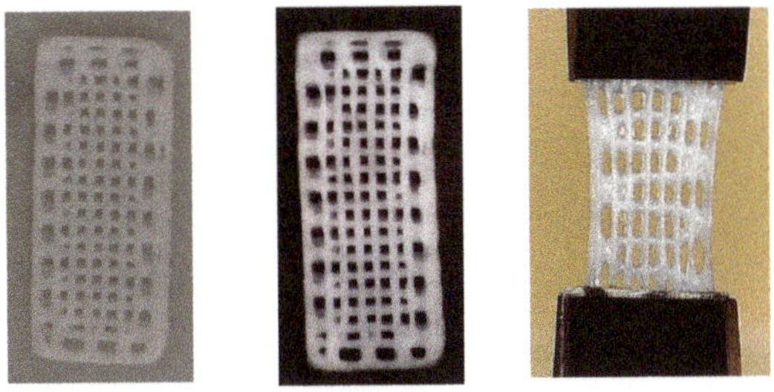
4L-3	4L-4
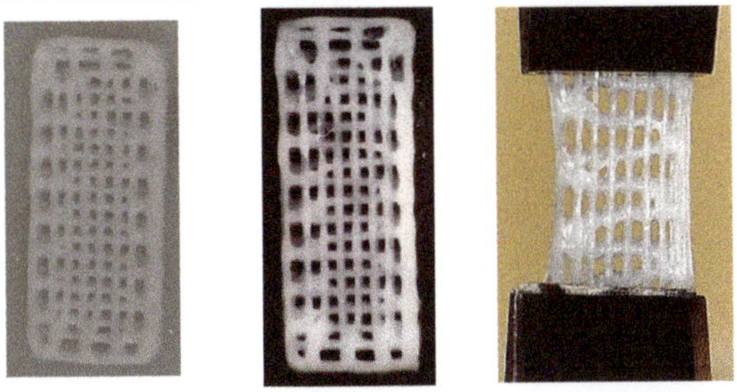	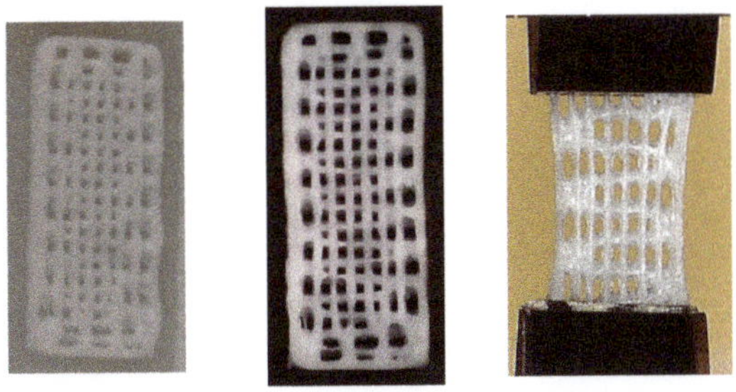
6L-1	6L-2
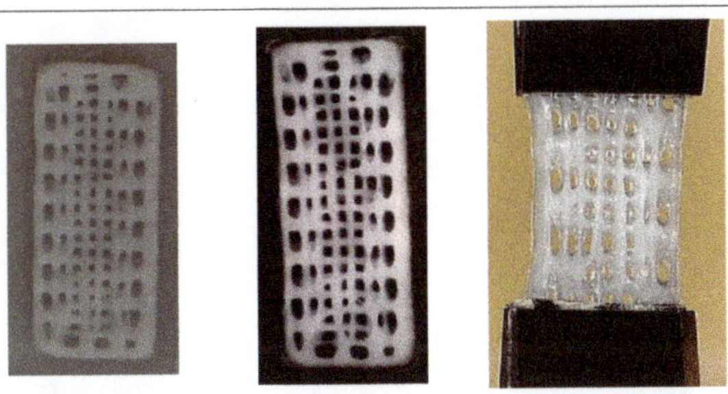	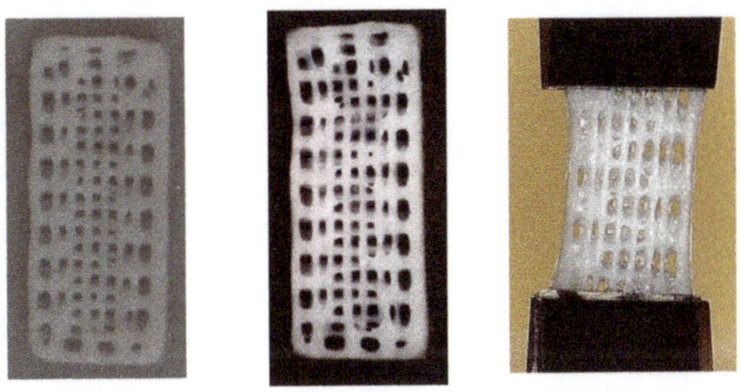
6L-3	6L-4
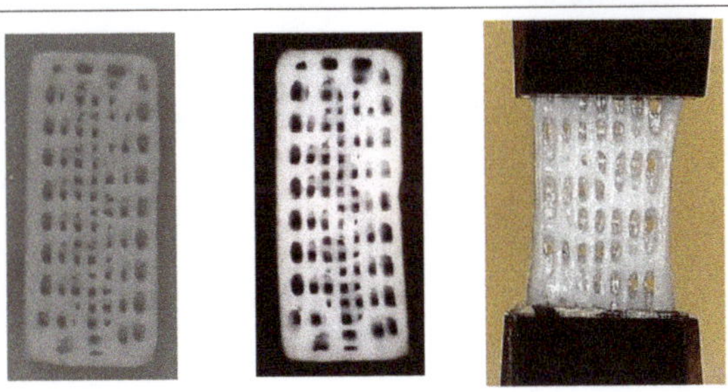	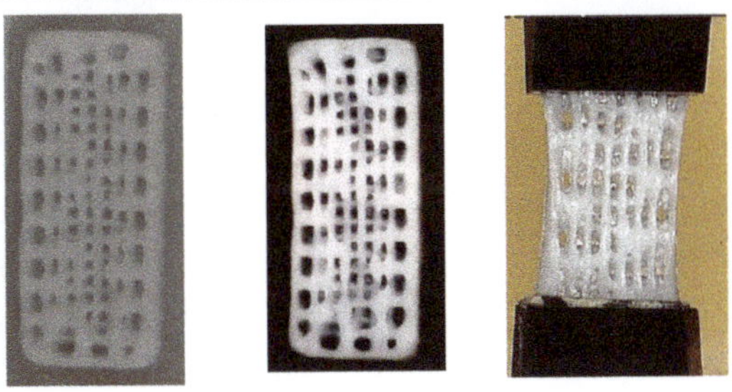

Appendix A.2. 3D Input Models for Sample Fabrication

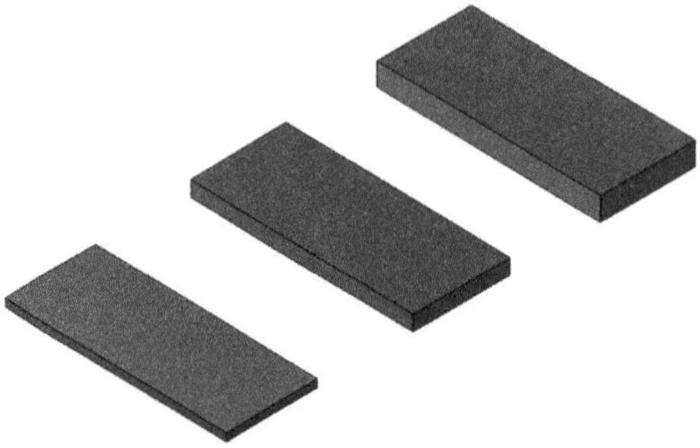

Figure A1. Three-dimensional CAD models of the three sample configurations used for fabrication. From left to right, the models correspond to the two-layer, four-layer, and six-layer PVA-C constructs. All samples share the same length and width (12.3 × 10 mm), while the thickness varies depending on the number of layers (1 mm, 2 mm, and 3 mm).

Appendix A.3. Sample Dimensions

Table A2. Width and height (in mm) of PVA-C samples with two, four, and six layers.

Sample	Width (mm)	Height
2L-1	12	0.73
2L-2	11.79	0.83
2L-3	12.49	0.66
2L-4	11.71	0.91
4L-1	12.28	1.8
4L-2	11.6	1.79
4L-3	11.26	1.79
4L-4	12.08	1.76
6L-1	12.41	2.36
6L-2	12.01	2.38
6L-3	11.65	2.44
6L-4	11.53	2.62

Appendix B. Mechanical Testing Parameters and Regression Analysis

Appendix B.1. DMA Testing Condition

Table A3. Testing condition file parameters for DMA of samples.

Condition Number	Frequency (Hz)	Mean Level (mm)	Mean Level Dwell (sec)	Dynamic Amplitude (mm)	Hold Value (mm)	Hold Value Dwell (sec)
1	1	3	2	0.4	0	0
2	2	3	2	0.4	0	0
3	3	3	2	0.4	0	0
4	4	3	2	0.4	0	0
5	6	3	2	0.4	0	0
6	8	3	2	0.4	0	0
7	10	3	2	0.4	0	0
8	1	3	2	0.4	0	0

Appendix B.2. Regression Model Coefficients

Table A4. Coefficient of determination (R^2) values for linear and quadratic fits across all samples and test repetitions.

Sample	Repetition	R^2 (Linear)	R^2 (Quadratic)
2L-1	T2	0.9947	0.9997
2L-1	T3	0.9948	0.9997
2L-1	T4	0.9957	0.9997
2L-1	T5	0.9964	0.9996
2L-2	T2	0.9962	0.9994
2L-2	T3	0.9961	0.9996
2L-2	T4	0.9962	0.9997
2L-2	T5	0.9962	0.9997
2L-3	T2	0.9974	0.9995
2L-3	T3	0.9965	0.9997
2L-3	T4	0.9963	0.9997
2L-3	T5	0.9965	0.9997
2L-4	T2	0.9918	0.9995
2L-4	T3	0.9908	0.9996
2L-4	T4	0.9899	0.9996
2L-4	T5	0.9897	0.9996
4L-1	T2	0.9995	0.9996
4L-1	T3	0.9976	0.9997
4L-1	T4	0.9945	0.9998
4L-1	T5	0.9937	0.9998
4L-2	T2	0.9973	0.9995
4L-2	T3	0.9967	0.9997
4L-2	T4	0.9964	0.9997
4L-2	T5	0.9963	0.9997
4L-3	T2	0.9919	0.9997
4L-3	T3	0.9906	0.9997
4L-3	T4	0.9903	0.9998
4L-3	T5	0.9903	0.9998
4L-4	T2	0.9981	0.9995
4L-4	T3	0.9971	0.9998
4L-4	T4	0.9968	0.9998
4L-4	T5	0.9972	0.9998
6L-1	T2	0.9985	0.9997
6L-1	T3	0.9981	0.9997
6L-1	T4	0.9984	0.9997
6L-1	T5	0.9985	0.9997
6L-2	T2	0.9972	0.9995
6L-2	T3	0.9951	0.9998
6L-2	T4	0.9957	0.9998
6L-2	T5	0.9960	0.9998
6L-3	T2	0.9987	0.9994
6L-3	T3	0.9984	0.9996
6L-3	T4	0.9989	0.9996
6L-3	T5	0.9989	0.9996
6L-4	T2	0.9942	0.9997
6L-4	T3	0.9940	0.9998
6L-4	T4	0.9946	0.9998
6L-4	T5	0.9943	0.9998

Appendix B.3. Young's Modulus and Variability Data

Table A5. Young's modulus and standard deviation across strain levels for 2L, 4L, and 6L samples.

Strain Levels (%)	Mean Young's Modulus, 2L (kPa)	Standard Deviation, 2L (kPa)	Mean Young's Modulus, 4L (kPa)	Standard Deviation, 4L (kPa)	Mean Young's Modulus, 6L (kPa)	Standard Deviation, 6L (kPa)
10	47.52	8.07	39.61	5.46	34.75	6.93
20	48.62	7.54	40.75	6.23	35.26	7.53
30	51.27	7.50	42.51	7.55	36.72	8.21
40	54.99	7.47	44.99	8.94	38.75	8.86

Appendix B.4. Raw DMA Modulus Data by Sample

Table A6. Individual storage (E′) and loss modulus (E″) values for each sample across 1–10 Hz.

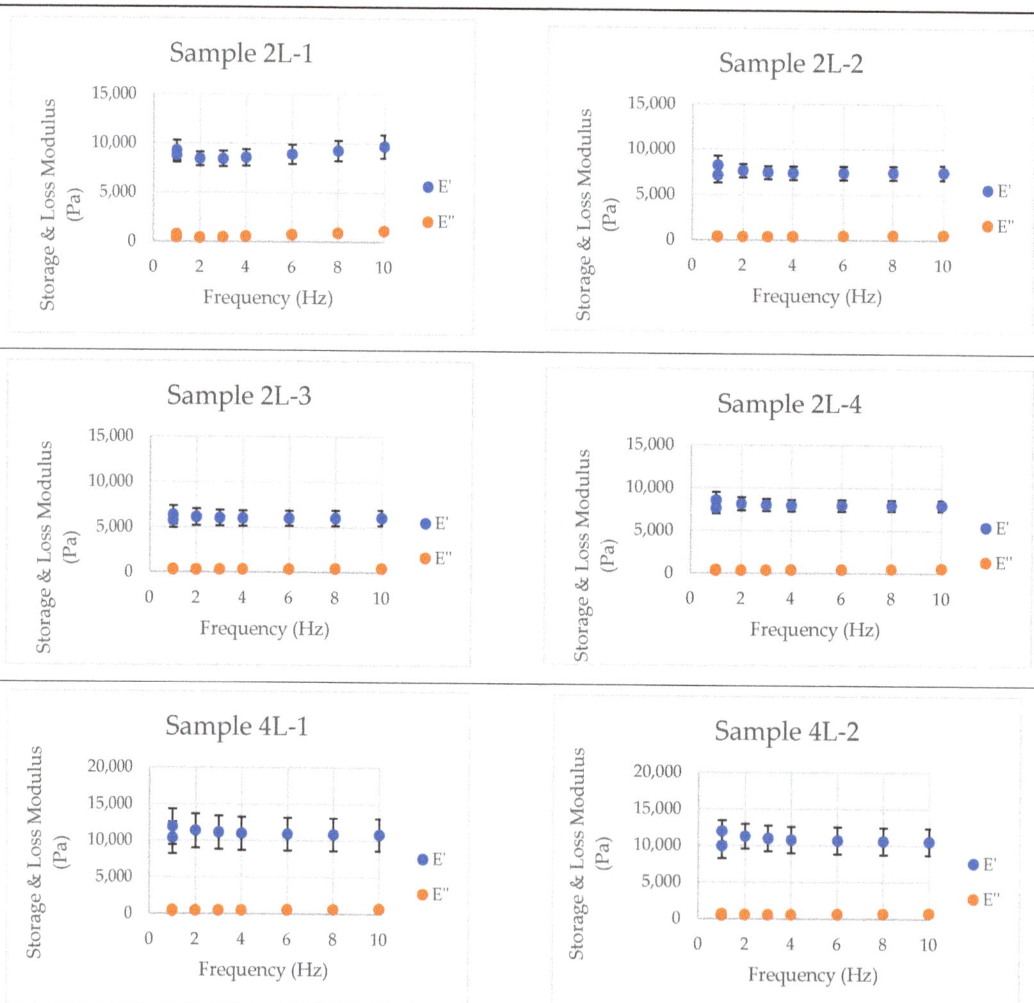

Table A6. *Cont.*

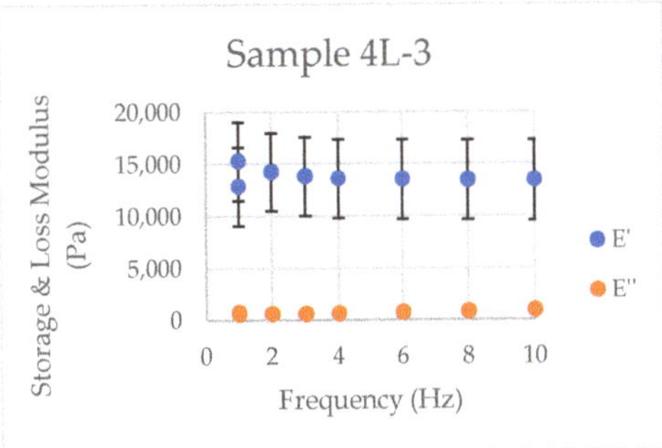

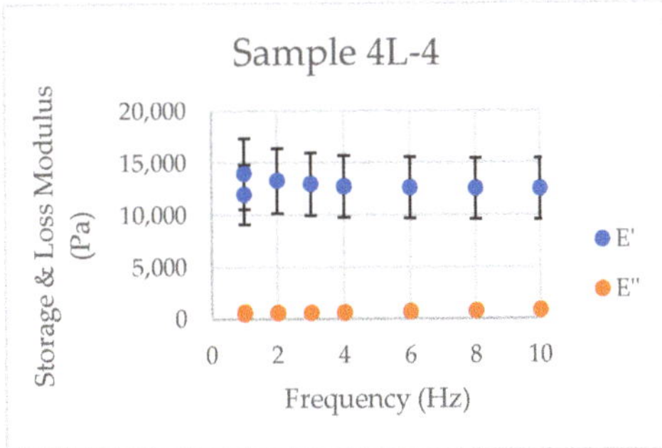

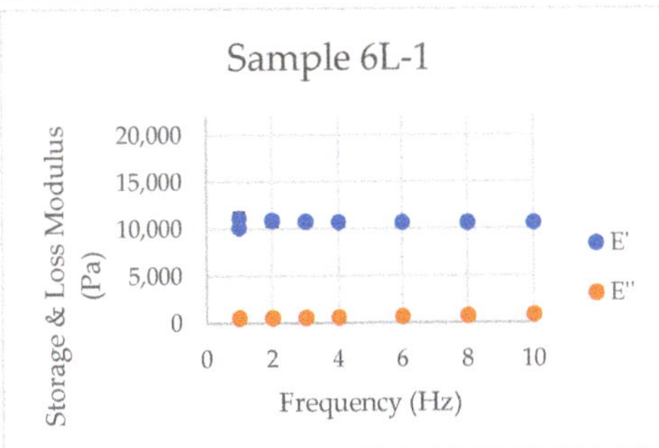

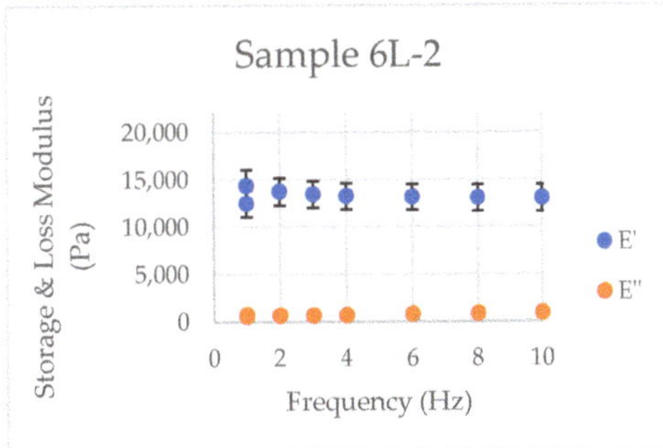

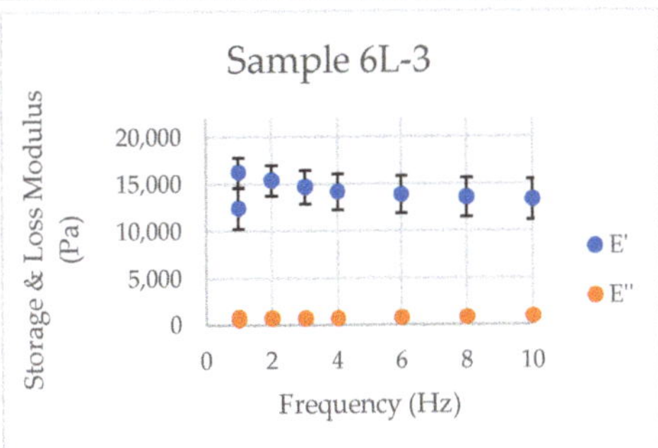

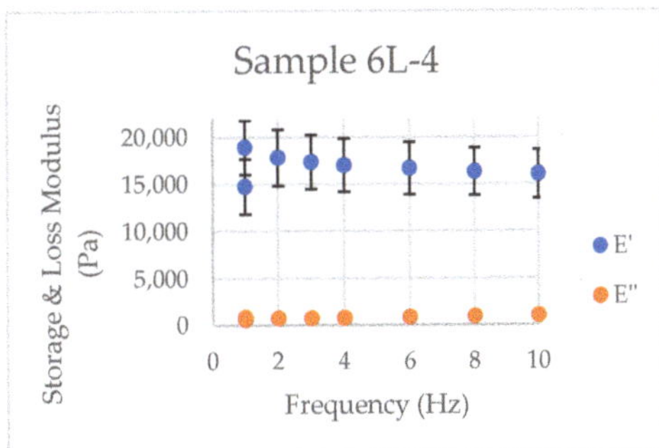

Appendix C. Statistical Analysis

Appendix C.1. Inter- and Intra-Sample Variability (ANOVA)

Table A7. Results from one-way ANOVA tests (95% confidence level) investigating inter-sample variability for YM at different strain levels.

Strain Level (%)	*p*-Value	Statistical Difference Between Sample Means
Inter-Sample Variability of Two-Layered (2L) Samples		
10	0.907	$p > 0.05$, no statistical difference
20	0.884	$p > 0.05$, no statistical difference
30	0.075	$p > 0.05$, no statistical difference
40	0.143	$p > 0.05$, no statistical difference

Table A7. *Cont.*

Strain Level (%)	p-Value	Statistical Difference Between Sample Means
Inter-Sample Variability of Four-Layered (4L) Samples		
10	0.018	$p < 0.05$, Sample 4L-2 different from 4L-1 and 4L-3
20	0.015	$p < 0.05$, Sample 4L-2 different from 4L-1 and 4L-3
30	0.137	$p > 0.05$, no statistical difference
40	0.029	$p < 0.05$, Sample 4L-2 different from 4L-1
Inter-Sample Variability of Six-Layered (6L) Samples		
10	0.008	$p < 0.05$, Sample 6L-1 different from 6L-4
20	0.008	$p < 0.05$, Sample 6L-1 different from 6L-4
30	0.069	$p > 0.05$, no statistical difference
40	0.147	$p > 0.05$, no statistical difference

Table A8. Results from one-way ANOVA tests (95% confidence level) investigating data linearity by checking for statistical differences between YM values at different strain levels (10%, 20%, 30%, 40%).

Sample Type	p-Value	Statistical Difference
2L	$p < 0.001$	$p < 0.05$, significant differences between strain levels
4L	$p = 0.0039$	$p < 0.05$, significant differences between strain levels
6L	$p = 0.0805$	$p > 0.05$, no significant differences between strain levels

Appendix C.2. Tukey's Post-Hoc Test Result

Table A9. Post-hoc one-way ANOVA results comparing YM at individual strain levels (10%, 20%, 30%, 40%).

Sample Type	p-Value	Statistical Difference
2L	<0.001	$p < 0.05$, 40% strain different from 10%, 20%, 30%
4L	<0.001	$p < 0.05$, 40% strain different from 10%, 20%
6L	0.798	$p > 0.05$, no statistical difference

Table A10. Tukey's multiple comparison test results for YM at 40% strain.

Comparison	Lower Bound	Mean Difference (Estimate)	Upper Bound	p-Value	Significance
2L vs. 4L	2.50	8.63	14.77	0.0039	Significant
2L vs. 6L	8.09	14.23	20.36	3.38×10^{-6}	Highly Significant
4L vs. 6L	−0.54	5.59	11.73	0.0805	Not Significant

Table A11. Results from one-way ANOVA tests (95% confidence level) for storage modulus at 2 Hz, 6 Hz, and 10 Hz.

Sample Type	Frequency (Hz)	p-Value	Statistical Difference
2L	2 Hz	0.002	$p < 0.05$, significant differences
4L	2 Hz	0.003	$p < 0.05$, significant differences
6L	2 Hz	0.005	$p < 0.05$, significant differences
2L	6 Hz	0.002	$p < 0.05$, significant differences
4L	6 Hz	0.003	$p < 0.05$, significant differences
6L	6 Hz	0.005	$p < 0.05$, significant differences
2L	10 Hz	0.002	$p < 0.05$, significant differences
4L	10 Hz	0.003	$p < 0.05$, significant differences
6L	10 Hz	0.005	$p < 0.05$, significant differences

Table A12. Results from one-way ANOVA tests (95% confidence level) for loss modulus at 2 Hz, 6 Hz, and 10 Hz.

Sample Type	Frequency (Hz)	p-Value	Statistical Difference
2L	2 Hz	0.002	$p < 0.05$, significant differences
4L	2 Hz	0.042	$p < 0.05$, significant differences
6L	2 Hz	0.300	$p > 0.05$, no statistical difference
2L	6 Hz	0.002	$p < 0.05$, significant differences
4L	6 Hz	0.042	$p < 0.05$, significant differences
6L	6 Hz	0.300	$p > 0.05$, no statistical difference
2L	10 Hz	0.002	$p < 0.05$, significant differences
4L	10 Hz	0.042	$p < 0.05$, significant differences
6L	10 Hz	0.300	$p > 0.05$, no statistical difference

Table A13. Results from Tukey's HSD storage modulus at 2 Hz.

Comparison	Mean Difference	p-Value (Adjusted)	Lower Bound	Upper Bound	Statistical Significance
2L vs. 4L	4950.1	0.016	1032.38	8867.819	Significant
2L vs. 6L	6854.267	0.002	2936.547	10,771.99	Significant
4L vs. 6L	1904.168	0.402	−2013.55	5821.887	Not Significant

Table A14. Results from Tukey's HSD storage modulus at 6 Hz.

Comparison	Mean Difference	p-Value (Adjusted)	Lower Bound	Upper Bound	Statistical Significance
2L vs. 4L	4368.505	0.018	832.893	7904.117	Significant
2L vs. 6L	6011.1	0.0027	2475.488	9546.712	Significant
4L vs. 6L	1642.595	0.4314	−1893.017	5178.207	Not Significant

Table A15. Results from Tukey's HSD storage modulus at 10 Hz.

Comparison	Mean Difference	p-Value (Adjusted)	Lower Bound	Upper Bound	Statistical Significance
2L vs. 4L	4025.866	0.025	560.388	7491.345	Significant
2L vs. 6L	5454.201	0.004	1988.722	8919.68	Significant
4L vs. 6L	1428.335	0.509	−2037.144	4893.814	Not Significant

Table A16. Results from Tukey's HSD loss modulus at 2 Hz.

Comparison	Mean Difference	p-Value (Adjusted)	Lower Bound	Upper Bound	Statistical Significance
2L vs. 4L	175.4693	0.0197	31.0644	319.8743	Significant
2L vs. 6L	270.3606	0.0014	125.9556	414.7656	Significant
4L vs. 6L	94.8912	0.2131	−49.5138	239.2962	Not Significant

Table A17. Results from Tukey's HSD loss modulus at 6 Hz.

Comparison	Mean Difference	p-Value (Adjusted)	Lower Bound	Upper Bound	Statistical Significance
2L vs. 4L	133.634	0.262	−87.581	354.848	Not Significant
2L vs. 6L	239.381	0.035	18.167	460.596	Significant
4L vs. 6L	105.748	0.413	−115.467	326.963	Not Significant

Table A18. Results from Tukey's HSD loss modulus at 10 Hz.

Comparison	Mean Difference	p-Value (Adjusted)	Lower Bound	Upper Bound	Statistical Significance
2L vs. 4L	107.092	0.716	−269.422	483.606	Not Significant
2L vs. 6L	223.610	0.272	−152.815	600.214	Not Significant
4L vs. 6L	116.608	0.675	−259.907	493.122	Not Significant

Appendix D. Dehydration Analysis

Appendix D.1. Sample Weight Change During DMA

Table A19. Sample weight changes before and after DMA testing.

Sample Number	Weight Before DMA (g) (Day 1, 1st DMA Test)	Weight After DMA (g) (Day 1, 1st DMA Test)	% Change (Day 1)	Weight Before DMA (g) (Day 2, 2nd DMA Test)	Weight After DMA (g) (Day 2, 2nd DMA Test)	% Change (Day 2)
2L-1	0.103	0.081	−21.63%	0.103	0.080	−22.08%
2L-2	0.131	0.108	−17.94%	0.126	0.103	−18.00%
2L-3	0.125	0.101	−18.88%	0.122	0.103	−15.53%
2L-4	0.139	0.114	−17.73%	0.134	0.114	−15.35%
4L-1	0.262	0.230	−12.28%	0.259	0.230	−11.23%
4L-2	0.267	0.245	−8.38%	0.264	0.236	−10.45%
4L-3	0.264	0.234	−11.18%	0.255	0.227	−11.10%
4L-4	0.274	0.246	−10.45%	0.279	0.237	−14.79%
6L-1	0.408	0.371	−8.99%	0.414	0.378	−8.51%
6L-2	0.408	0.374	−8.21%	0.413	0.377	−8.69%
6L-3	0.422	0.371	−12.08%	0.432	0.389	−10.00%
6L-4	0.419	0.378	−9.67%	0.416	0.377	−9.37%

References

1. British Heart Foundation. *Global Heart & Circulatory Diseases Factsheet*; British Heart Foundation: London, UK, 2024; pp. 1–12. Available online: https://www.bhf.org.uk/-/media/files/for-professionals/research/heart-statistics/bhf-cvd-statistics-global-factsheet.pdf (accessed on 27 March 2025).
2. Venkatraman, S.; Boey, F.; Lao, L.L. Implanted cardiovascular polymers: Natural, synthetic and bio-inspired. *Prog. Polym. Sci.* **2008**, *33*, 853–874. [CrossRef]
3. Fukuchi, K. Proposal and feasibility study of a 6-level PSK modulation format based system for 100-Gb/s migration. In Proceedings of the OFC/NFOEC 2008—2008 Conference on Optical Fiber Communication/National Fiber Optic Engineers Conference, San Diego, CA, USA, 24–28 February 2008; Volume 35, pp. 348–355.
4. Holzapfel, G.A.; Ogden, R.W. Modelling the layer-specific three-dimensional residual stresses in arteries, with an application to the human aorta. *J. R. Soc. Interface* **2010**, *7*, 787–799. [CrossRef] [PubMed]
5. Balint, B.; Jepchumba, V.K.; Guéant, J.L.; Guéant-Rodriguez, R.M. Mechanisms of homocysteine-induced damage to the endothelial, medial and adventitial layers of the arterial wall. *Biochimie* **2020**, *173*, 100–106. [CrossRef]
6. Rafuse, M.; Xu, X.; Stenmark, K.; Neu, C.P.; Yin, X.; Tan, W. Layer-specific arterial micromechanics and microstructure: Influences of age, anatomical location, and processing technique. *J. Biomech.* **2019**, *88*, 113–121. [CrossRef] [PubMed]
7. Lo Vercio, L.; del Fresno, M.; Larrabide, I. Lumen-intima and media-adventitia segmentation in IVUS images using supervised classifications of arterial layers and morphological structures. *Comput. Methods Programs Biomed.* **2019**, *177*, 113–121. [CrossRef]
8. Wagenseil, J.E.; Mecham, R.P. Vascular extracellular matrix and arterial mechanics. *Physiol. Rev.* **2009**, *89*, 957–989. [CrossRef]
9. Greco Song, H.-H.; Rumma, R.T.; Ozaki, C.K.; Edelman, E.R.; Chen, C.S. Vascular tissue engineering: Progress, challenges, and clinical promise. *Cell Stem Cell* **2018**, *22*, 608. [CrossRef]
10. Khaniki, H.B.; Ghayesh, M.H.; Chin, R.; Amabili, M. Hyperelastic structures: A review on the mechanics and biomechanics. *Int. J. Non. Linear. Mech.* **2023**, *148*, 104275. [CrossRef]
11. Ho, T.C.; Chang, C.C.; Chan, H.P.; Chung, T.W.; Shu, C.W.; Chuang, K.P.; Duh, T.H.; Yang, M.H.; Tyan, Y.C. Hydrogels: Properties and applications in biomedicine. *Molecules* **2022**, *27*, 2902. [CrossRef]

12. Oztemur, J.; Ozdemir, S.; Sezgin, H.; Yalcin-Enis, I. *Hydrogel-Based Vascular Grafts: State of Art*; Elsevier Inc.: Amsterdam, The Netherlands, 2023.
13. Gupta, S.; Goswami, S.; Sinha, A. A combined effect of freeze-thaw cycles and polymer concentration on the structure and mechanical properties of transparent PVA gels. *Biomed. Mater.* **2012**, *7*, 015006. [CrossRef]
14. Kantaros, A. 3D printing in regenerative medicine: Technologies and resources utilized. *Int. J. Mol. Sci.* **2022**, *23*, 14621. [CrossRef]
15. Kantaros, A.; Ganetsos, T.; Petrescu, F.I.T.; Alysandratou, E. Bioprinting and intellectual property: Challenges, opportunities, and the road ahead. *Bioengineering* **2025**, *12*, 76. [CrossRef]
16. Liu, Y.; Geever, L.M.; Kennedy, J.E.; Higginbotham, C.L.; Cahill, P.A.; McGuinness, G.B. Thermal behavior and mechanical properties of physically crosslinked PVA/Gelatin hydrogels. *J. Mech. Behav. Biomed. Mater.* **2010**, *3*, 203–209. [CrossRef] [PubMed]
17. Gasser, T.C.; Ogden, R.W.; Holzapfel, G.A. Hyperelastic modelling of arterial layers with distributed collagen fibre orientations. *J. R. Soc. Interface* **2006**, *3*, 15–35. [CrossRef]
18. Fleischer, S.; Tavakol, D.N.; Vunjak-Novakovic, G. From arteries to capillaries: Approaches to engineering human vasculature. *Adv. Funct. Mater.* **2020**, *30*, 190811. [CrossRef]
19. Mihaylova, A.; Shopova, D.; Parahuleva, N.; Yaneva, A.; Bakova, D. (3D) Bioprinting—Next dimension of the pharmaceutical sector. *Pharmaceuticals* **2024**, *17*, 797. [CrossRef]
20. Yaneva, A.; Shopova, D.; Bakova, D.; Mihaylova, A.; Kasnakova, P.; Hristozova, M.; Semerdjieva, M. The progress in bioprinting and its potential impact on health-related quality of life. *Bioengineering* **2023**, *10*, 910. [CrossRef]
21. Christensen, K.; Davis, B.; Jin, Y.; Huang, Y. Effects of printing-induced interfaces on localized strain within 3D printed hydrogel structures. *Mater. Sci. Eng. C* **2018**, *89*, 65–74. [CrossRef]
22. Giudici, A.; Khir, A.W.; Szafron, J.M.; Spronck, B. From uniaxial testing of isolated layers to a tri-layered arterial wall: A novel constitutive modelling framework. *Ann. Biomed. Eng.* **2021**, *49*, 2454–2467. [CrossRef]
23. Pazos, V.; Mongrain, R.; Tardif, J.C. Polyvinyl alcohol cryogel: Optimizing the parameters of cryogenic treatment using hyperelastic models. *J. Mech. Behav. Biomed. Mater.* **2009**, *2*, 542–549. [CrossRef]
24. Webb, B.; Doyle, B.J. Parameter optimization for 3D bioprinting of hydrogels. *Bioprinting* **2017**, *8*, 8–12. [CrossRef]
25. Bento, C.S.A.; Gaspar, M.C.; Coimbra, P.; de Sousa, H.C.; Braga, M.E.M. A review of conventional and emerging technologies for hydrogels sterilization. *Int. J. Pharm.* **2023**, *634*, 122671. [CrossRef] [PubMed]
26. Adelnia, H.; Ensandoost, R.; Moonshi, S.S.; Gavgani, J.N.; Vasafi, E.I.; Ta, H.T. Freeze/thawed polyvinyl alcohol hydrogels: Present, past and future. *Eur. Polym. J.* **2022**, *164*, 110974. [CrossRef]
27. Fegan, K.L. Engineering the Biomechanics of PVA/Gelatin Cryogels for Coronary Artery. Ph.D. Thesis, University of Birmingham, Birmingham, UK, 2023.
28. Crolla, J.P. Replicating the Intrinsic Biomechanical Characteristics of Connective Tissue: An Exploration into the Design and Manufacture of PVA Cryogel. Ph.D. Thesis, University of Birmingham, Birmingham, UK, 2022.
29. Burton, H.E.; Freij, J.M.; Espino, D.M. Dynamic viscoelasticity and surface properties of porcine left anterior descending coronary arteries. *Cardiovasc. Eng. Technol.* **2017**, *8*, 41–56. [CrossRef]
30. Patino-Alonso, M.C.; Molina, J.L.; Zazo, S. Multivariate linear modeling for the application in the field of hydrological engineering. In *Handbook of HydroInformatics Vol. III: Water Data Management Best Practices*; Elsevier: Amsterdam, The Netherlands, 2022; pp. 277–289. [CrossRef]
31. Einav, S.; O'Connor, M. P-values and significance: The null hypothesis that they are not related is correct. *J. Crit. Care* **2019**, *54*, 159–162. [CrossRef]
32. Kim, H.-Y. Statistical notes for clinical researchers: Effect size. *Restor. Dent. Endod.* **2015**, *40*, 328. [CrossRef]
33. Groll, J.; Burdick, J.A.; Cho, D.W.; Derby, B.; Gelinsky, M.; Heilshorn, S.C.; Juengst, T.; Malda, J.; Mironov, V.A.; Nakayama, K.; et al. A definition of bioinks and their distinction from biomaterial inks. *Biofabrication* **2019**, *11*, 013001. [CrossRef]
34. Fung, Y.C. *Biomechanics: Mechanical Properties of Living Tissues*; Springer: New York, NY, USA, 1993.
35. Ali, A.; Hosseini, M.; Sahari, B.B. A review of constitutive models for rubber-like materials. *Am. J. Eng. Appl. Sci.* **2010**, *3*, 232–239. [CrossRef]
36. Crolla, J.P.; Britton, M.M.; Espino, D.M.; Thomas-Seale, L.E.J. The dynamic viscoelastic characterisation and magnetic resonance imaging of poly(vinyl alcohol) cryogel: Identifying new attributes and opportunities. *Mater. Sci. Eng. C* **2021**, *129*, 112383. [CrossRef] [PubMed]
37. Gale, L.; Panieraki, A.; Mahmoodi, N.; Crolla, J.P.; Thomas-Seale, L.E.J. The design and characterisation of sinusoidal toolpaths using sub-zero bioprinting of polyvinyl alcohol. *J. Mech. Behav. Biomed. Mater.* **2024**, *152*, 106402. [CrossRef]
38. Dutta, D.; Lee, K.W.; Allen, R.A.; Wang, Y.; Brigham, J.C.; Kim, K. Non-invasive assessment of elastic modulus of arterial constructs during cell culture using ultrasound elasticity imaging. *Ultrasound Med. Biol.* **2013**, *39*, 2103–2115. [CrossRef] [PubMed]
39. Holzapfel, G.A.; Sommer, G.; Gasser, C.T.; Regitnig, P. Determination of layer-specific mechanical properties of human coronary arteries with nonatherosclerotic intimal thickening and related constitutive modeling. *Am. J. Physiol. Heart Circ. Physiol.* **2005**, *289*, H2048–H2058. [CrossRef] [PubMed]

40. Lim, J.; Won, J.Y.; Ahn, C.B.; Kim, J.; Kim, H.J.; Jung, J.S. Comparison of hemodynamic energy between expanded polytetrafluoroethylene and Dacron artificial vessels. *J. Chest Surg.* **2021**, *54*, 81–87. [CrossRef]
41. Rodríguez-Soto, M.A.; Polanía-Sandoval, C.A.; Aragón-Rivera, A.M.; Buitrago, D.; Ayala-Velásquez, M.; Velandia-Sánchez, A.; Peralta Peluffo, G.; Cruz, J.C.; Muñoz Camargo, C.; Camacho-Mackenzie, J.; et al. Small diameter cell-free tissue-engineered vascular grafts: Biomaterials and manufacture techniques to reach suitable. *Polymers* **2022**, *14*, 3440. [CrossRef]
42. Liu, Y.; Zahedmanesh, H.; Lally, C.; Cahill, P.A.; McGuinness, G.B. Compliance properties of a composite electrospun fibre-hydrogel blood vessel scaffold. *Mater. Lett.* **2016**, *178*, 296–299. [CrossRef]
43. Miramini, S.; Fegan, K.L.; Green, N.C.; Espino, D.M.; Zhang, L.; Thomas-Seale, L.E.J. The status and challenges of replicating the mechanical properties of connective tissues using additive manufacturing. *J. Mech. Behav. Biomed. Mater.* **2020**, *103*, 103544. [CrossRef]
44. Li, M.G.; Tian, X.Y.; Chen, X.B. Modeling of flow rate, pore size, and porosity for the dispensing-based tissue scaffolds fabrication. *J. Manuf. Sci. Eng.* **2009**, *131*, 0345011–0345015. [CrossRef]
45. Li, Y.; Hu, C.; Lan, J.; Yan, B.; Zhang, Y.; Shi, L.; Ran, R. Hydrogel-based temperature sensor with water retention, frost resistance and remoldability. *Polymer* **2020**, *186*, 122027. [CrossRef]
46. Crolla, J.P.; Britton, M.M.; Espino, D.M.; Thomas-Seale, L.E.J. The orthotropic viscoelastic characterisation of sub-zero 3D-printed poly(vinyl alcohol) cryogel. *MRS Adv.* **2021**, *6*, 467–471. [CrossRef]
47. Paxton, N.; Smolan, W.; Böck, T.; Melchels, F.; Groll, J.; Jungst, T. Proposal to assess printability of bioinks for extrusion-based bioprinting and evaluation of rheological properties governing bioprintability. *Biofabrication* **2017**, *9*, 044107. [CrossRef]
48. Plugariu, I.A.; Bercea, M. The viscosity of globular proteins in the presence of an 'inert' macromolecular cosolute. *J. Mol. Liq.* **2021**, *337*, 116382. [CrossRef]
49. Chuong, C.J.; Fung, Y.C. Biomechanical Engineering DE. *J. Biomech. Eng.* **1986**, *108*, 189–192. [CrossRef] [PubMed]
50. Wang, M.; Bai, J.; Shao, K.; Tang, W.; Zhao, X.; Lin, D.; Huang, S.; Chen, C.; Ding, Z.; Ye, J. Poly(vinyl alcohol) hydrogels: The old and new functional materials. *Int. J. Polym. Sci.* **2021**, *2021*, 2225426. [CrossRef]
51. Hassan, C.M.; Peppas, N.A. Structure and morphology of freeze/thawed PVA hydrogels. *Macromolecules* **2000**, *33*, 2472–2479. [CrossRef]
52. Daza Agudelo, J.I.; Ramirez, M.R.; Henquin, E.R.; Rintoul, I. Modelling of swelling of PVA hydrogels considering non-ideal mixing behaviour of PVA and water. *J. Mater. Chem. B* **2019**, *7*, 4049–4054. [CrossRef]
53. Sun, Z.; Song, C.; Wang, C.; Hu, Y.; Wu, J. Hydrogel-based controlled drug delivery for cancer treatment: A review. *Mol. Pharm.* **2020**, *17*, 373–391. [CrossRef]
54. Bialik-Wąs, K.; Pluta, K.; Malina, D.; Majka, T.M. Alginate/PVA-based hydrogel matrices with Echinacea purpurea extract as a new approach to dermal wound healing. *Int. J. Polym. Mater. Polym. Biomater.* **2021**, *70*, 195–206. [CrossRef]
55. Chen, C.; Zhu, Y.; Bao, H.; Zhao, P.; Jiang, H.; Peng, L.; Yang, X.; Li, C. Solvent-assisted poly(vinyl alcohol) gelated crystalline colloidal array photonic crystals. *Soft Matter* **2011**, *7*, 915–921. [CrossRef]
56. Yang, Z.; Peng, H.; Wang, W.; Liu, T. Crystallization behavior of poly(ε-caprolactone)/layered double hydroxide nanocomposites. *J. Appl. Polym. Sci.* **2010**, *116*, 2658–2667. [CrossRef]
57. Masri, S.; Maarof, M.; Aziz, I.A.; Idrus, R.; Fauzi, M.B. Performance of hybrid gelatin-PVA bioinks integrated with genipin through extrusion-based 3D bioprinting: An in vitro evaluation using human dermal fibroblasts. *Int. J. Bioprint.* **2023**, *9*, 3. [CrossRef]
58. Hyun, G.W.; Park, S.H. 3D freeform fabrication of fiber-reinforced artificial vascular grafts with biomimetic softness and enhanced mechanical strength. *J. Manuf. Process.* **2025**, *133*, 1237–1248. [CrossRef]
59. Dell, A.C.; Maresca, J.; Davis, B.A.; Isaji, T.; Dardik, A.; Geibel, J.P. Development and deployment of a functional 3D-bioprinted blood vessel. *Sci. Rep.* **2025**, *15*, 11668. [CrossRef] [PubMed]
60. Reeser, K.; Doiron, A.L. Three-dimensional printing on a rotating cylindrical mandrel: A review of additive-lathe 3D printing technology. *3D Print. Addit. Manuf.* **2019**, *6*, 293–307. [CrossRef]
61. Rafat, M.; Rotenstein, L.S.; You, J.O.; Auguste, D.T. Dual functionalized PVA hydrogels that adhere endothelial cells synergistically. *Biomaterials* **2012**, *33*, 3880–3886. [CrossRef]
62. Cutiongco, M.F.A.; Goh, S.H.; Aid-Launais, R.; Le Visage, C.; Low, H.Y.; Yim, E.K.F. Planar and tubular patterning of micro and nano-topographies on poly(vinyl alcohol) hydrogel for improved endothelial cell responses. *Biomaterials* **2016**, *84*, 184–195. [CrossRef]
63. Chong, S.F.; Smith, A.A.A.; Zelikin, A.N. Microstructured, functional PVA hydrogels through bioconjugation with oligopeptides under physiological conditions. *Small* **2013**, *9*, 942–950. [CrossRef]

Journal of
Manufacturing and
Materials Processing

MDPI

Article

Permea-Design: An Innovative Tool for Generating Triply Periodic Minimal Surface Scaffolds with Tailored Permeability

Matthew Bedding-Tyrrell [1,2], Bjornar Sandnes [3], Perumal Nithiarasu [2] and Feihu Zhao [1,2,*]

[1] Department of Biomedical Engineering, Faculty of Science and Engineering, Swansea University, Swansea SA1 8EN, UK; m.j.a.bedding@swansea.ac.uk
[2] Zienkiewicz Institute for Modelling, Data and AI, Swansea University, Swansea SA1 8EN, UK; p.nithiarasu@swansea.ac.uk
[3] Complex Flow Lab, Department of Chemical Engineering, Faculty of Science and Engineering, Swansea University, Swansea SA1 8EN, UK; b.sandnes@swansea.ac.uk
* Correspondence: feihu.zhao@swansea.ac.uk; Tel.: +44-(0)-1792-606638

Abstract: The permeability of a porous material is the measure of the ability of fluids to pass through it. The ability to control permeability is valued by tissue engineers who manufacture tissue engineering scaffolds that house cells/tissue and facilitate tissue growth. Therefore, a scaffold design software in which permeability can be entered as a variable in determining the structure and strut topology would be a desirable tool for tissue engineering researchers. The ability to factor permeability directly into the design of scaffolds facilitates more effective bone tissue engineering by enabling optimal nutrient transport and waste removal at regeneration sites. Additionally, having the ability to control the mechanical environment by indicating a region of acceptable porosities for in vitro cell culturing is desirable. This desirability is a result of porosity being a major determining factor in permeability, where increasing porosity will generally mean a higher permeability. Thus, having an upper bound on porosity means that higher-permeability structures can be determined whilst maintaining high values of mechanical strength. In this software, a method is discussed for modifying the Kozeny–Carman equation by incorporating level-set equations for different triply periodic minimal surface (TPMS) structures. Topology analysis is computed on six different TPMS structures in the toolbox, and a relationship between a topological constant and permeability is derived through the Kozeny–Carman equation. This relationship allows for an input of permeability as a factor in the determination of pore size, porosity, and scaffold structure. This novel method allows for scaffold design based on a tailored permeability to assist successful tissue engineering.

Keywords: permeability; tissue engineering scaffold; TPMS structures; CAD toolbox

1. Introduction

Three-dimensional (3D) porous scaffolds play an important role in tissue engineering because they house the tissue cells, facilitate the growth of those cells, and produce the ECM (extracellular matrix) that is needed for tissue morphogenesis, differentiation, and homeostasis [1,2]. Among different scaffold designs, TPMS (triply periodic minimal surface) structures are valuable to bone tissue engineers because they combine an ability to mimic biological bone tissue with high mechanical strength, which is required for bone regeneration [3]. TPMS structures exhibit lower stress concentrations, which decreases the probability of scaffold failure whilst undergoing bone regeneration [4]. Many of the TPMS structures also have a high surface area to volume ratio, which allows for the optimal delivery of nutrients and oxygen to artificial bone tissues and regeneration sites [5].

Permeability is a key factor in influencing the environment of cells. It is the measure of how well a scaffold can direct and conduct the fluid flow. Fluid flow is important for two main reasons: it induces mechanical stimulation on cells, and it enables the delivery of nutrients to the growing cells [6]. Therefore, a higher permeability can enhance nutrient transport throughout the scaffold. Cell migration and growth during cell culture process is also dependent on the ability of nutrients to permeate through the scaffold [7]. The ability to tailor the scaffold permeability is valuable because it enables the adjustment of nutrient transport and mechanical stimulation on cells, which facilitates optimal growth conditions. It has previously been shown that scaffold porous geometries can influence the scaffold permeability and internal mechanical/fluidic environment within a bioreactor [8]. It has also been shown that tortuosity, porosity, and constriction factor all affect permeability [9]. A higher porosity will generally mean a higher permeability and lower wall shear stress under constant flow rate in a perfusion bioreactor. TPMS structures are known to have a high permeability, which allows for high levels of nutrient transport and waste removal. One important component in the high permeability of TPMS scaffolds is their low constriction factor, which is due to their periodicity in three independent directions. Another component of their high permeability is their low tortuosity, which is due to their network of interconnected pores. Thus, lower values of tortuosity and the constriction factor led to higher permeabilities [9]. Additional research has indicated permeability can also be described by porosity, specific surface area, and a geometric shape factor [10]. Software capable of estimating permeability is already widely used. For example, Alvascience is a software suite that can predict the permeability of the blood–brain barrier to an 'accuracy equal or greater than 0.8' [11]. Convolutional neural networks have also been used to predict permeability, where the permeability of low-resolution porous media were predicted to an R^2 value of 0.896, where R^2 is the coefficient of determination [12]. However, to the authors' knowledge, there is no currently available software that can generate a structure to a tailored permeability.

In this study, we aim to create a toolbox that can design scaffolds to a tailored permeability. To achieve this, it is hypothesized that the Kozeny–Carman equation, which was originally used for calculating the permeability of porous media, can be modified by incorporating the level-set constant from TPMS level-set equations [13]. To test this hypothesis, relationships that relate the level-set constant to porosity and specific surface area need to be derived and then integrated into the Kozeny–Carman equation.

2. Materials and Methods

2.1. Topology Analysis

A first step in modifying the Kozeny–Carman equation by incorporating the level-set constant is to complete topology analysis on each of the TPMS structures. Relationships that relate porosity, pore size, and specific surface area to the level-set constant need to be determined. The 6 structures used in this study are the Gyroid, FKS, Primitive, Diamond, FRD and Neovius [14], whose topologies are shown in Figure 1a. FKS (Fischer Koch S) and F-RD (F-rhombic dodecahedra) are two of the lesser known TPMS structures; however, they have the two highest specific surface areas out of the 6 TPMS structures. Therefore, it is hypothesized that they will have high value as tissue engineering scaffolds [15]. TPMS structures can take one of two different forms of lattice: a ligament-based lattice and a sheet-based lattice. In a ligament-based lattice, the surface separates two volumes. One of these volumes is solidified and makes up the scaffold; the other volume makes up the void space of the scaffold. In a sheet-based lattice, two surface equations are used with different level-set constant values. The gap between these two surfaces is solidified, and this is used to make up the scaffold [16]. A previous study showed close similarity between experimental and simulation results for ligament-based TPMS lattices, whilst for sheet-based lattices, there was significant divergence between the study's experimental and simulation results [17]. Therefore, in this study, we use ligament-based lattices for creating the TPMS geometries.

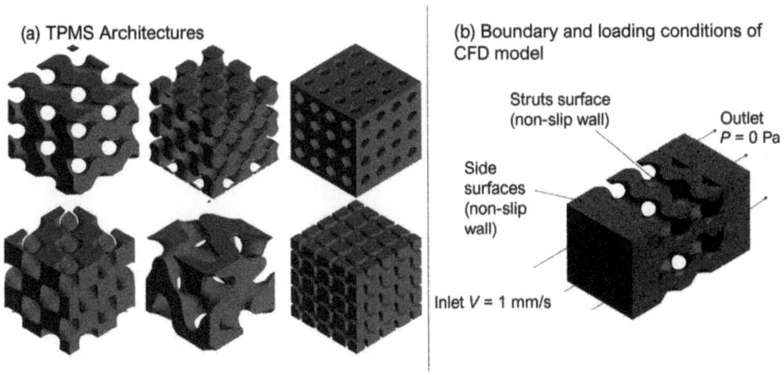

Figure 1. (**a**) Six TPMS architectures: gyroid, F-RD, primitive, diamond, FKS, Neovius (from top left to bottom right). (**b**) CFD simulation setup to determine Kozeny constant β.

One method of generating these surfaces is to use level-set approximation equations, based on trigonometric functions. The gyroid surface is approximated by the following equation [18]:

$$\Phi_G(x,y,z) = cos(x)sin(y) + cos(y)sin(z) + cos(z)sin(x) - C, \quad (1)$$

where C is the level-set constant and calculates the surface when $\Phi_G = 0$. The level-set constant controls the pore morphology and therefore can be altered to change the initial pore size and porosity [19].

The level-set approximation is controlled by both the trigonometric functions and the level-set constant. By calculating the level-set equation (Equation (1)) at a given Cartesian coordinate, it can be determined whether that point is in the solid or void phase. If the sum of the trigonometric functions is greater than C, then $\Phi > 0$, and the point is considered to be in the solid volume. If the sum is less than the value of C, then $\Phi < 0$ and the point lies within the void space. As the value of the level-set constant is altered, the morphology of the surface changes, thus changing the ratio of coordinates on each side of the surface. This shows that there is a direct relationship between the level-set constant and the porosity of a TPMS scaffold. This relationship has already been estimated for the gyroid surface (Equation (2)) [20]. In a similar method, altering the level-set constant will change the morphology of the surface, changing the pore size of the structure. This relationship has also been identified for the gyroid surface (Equation (3)) [20]:

$$C = 0.7864\varphi^3 - 1.1798\varphi^2 - 2.5259\varphi + 1.4597, \quad (2)$$

$$d_p{}^{2\pi} = -11.7311C^5 - 0.1307C^4 - 1.7987C^3 + 0.2070C^2 - 186.9928 + 433.0114, \quad (3)$$

where φ is porosity and $d_p{}^{2\pi}$ is the pore size of a structure prior to scaling. A structure can then be scaled to a specified pore size (d_p) using the ratio of d_p and $d_p{}^{2\pi}$.

Using a similar automatic method to Reference [20], level-set relationships were determined for the porosity, pore size, and unscaled specific surface area of each of the TPMS scaffolds. These relationships, along with the definitions of pore size for each TPMS structure, are given in the Supplementary Material. Specific surface area is a component of the Kozeny–Carman equation and is characterized as the wetted surface area of the scaffold divided by the bulk volume of the scaffold. Using these relationships and the Kozeny–Carman equation, code was written to test different $4\pi \times 4\pi \times 4\pi$ mm scaffolds to test how the level-set constant affects the geometric permeability. This scaffold size was chosen due to a volume dependency study to negate the boundary effects.

2.2. CFD Simulation to Determine Kozeny Constant β

To assess the accuracy of the Kozeny–Carman equation (Equation (5)) and the value of the Kozeny constant, β, 16 scaffolds were simulated using ANSYS CFX 2023 R2 (ANSYS Inc., Canonsburg, PA, USA) for each of the six TPMS architectures. The scaffolds ranged in porosity from 50 to 80% with 10% intervals, and for each porosity level there was a range of pore sizes from 400 to 700 μm with 100 μm intervals—this combination of four levels of porosity and four values for pore size giving 16 scaffolds for each architecture. The geometric inverse of the TPMS structures was extracted and used as the fluidic domain. The simulation setup was designed with a free fluid domain on opposite sides of each TPMS structure with an inlet velocity of 1 mm/s and a zero-pressure outlet applied to the end faces of the free-fluid domain with water as the medium. Non-slip walls were used on the other faces of the fluidic domain (Figure 1b). The fluid domains were imported as STL meshes, then regularized to a size of 50 μm, and then solidified (Figure 1b). Pressure–velocity coupling was used in CFX and was implemented using the Rhie–Chow algorithm [21]. The convergence criteria of root-mean-square residual of the mass and momentum is $<1 \times 10^{-4}$. All of the simulations were run on a computer with 128GB RAM and an Intel i7 12,700 CPU (12 cores).

The pressure drop was recorded from the inlet to the outlet of the scaffold region, and Darcy's law was used to determine the absolute permeability. The absolute permeability was then divided by the geometric permeability determined from the Kozeny–Carman Equation (5) to calculate the Kozeny constant, β, for each of the individual scaffolds.

2.3. Development of Scaffold Design Toolbox

The toolbox functions following the flowchart seen in Figure 2. It uses the updated Kozeny–Carman equation to return values for porosity and pore size for each of the TPMS types. The ψ–level-set constant relationship (Equation (8)) is used to find values of pore size for input values of permeability, β, and porosity. This relationship is derived automatically within the toolbox for all 6 of the TPMS types. The user will be required to input a permeability and set the ranges for porosity and pore size. As can be seen in Figure 2, the toolbox will test 6 different porosities within the specified porosity range. The level-set constant of each of the different porosities is determined using Equation (2); the gyroid relationship is shown graphically in Figure 3b. The level-set constant (C) is then used to determine ψ using the equation from the fitting functions in Figure 4. Using permeability, ψ, and β, pore size can be determined using Equation (9). If any of the 6 tested porosities produces a pore size within the pore size range, the porosity, pore size, and TPMS type are recorded, and a scaffold is generated and then can be exported. If no pore sizes are found to be in the specified range, the program will run again with a different TPMS type. This process will repeat until an acceptable pore size is found or until all 6 TPMS types have been tested with a nil return.

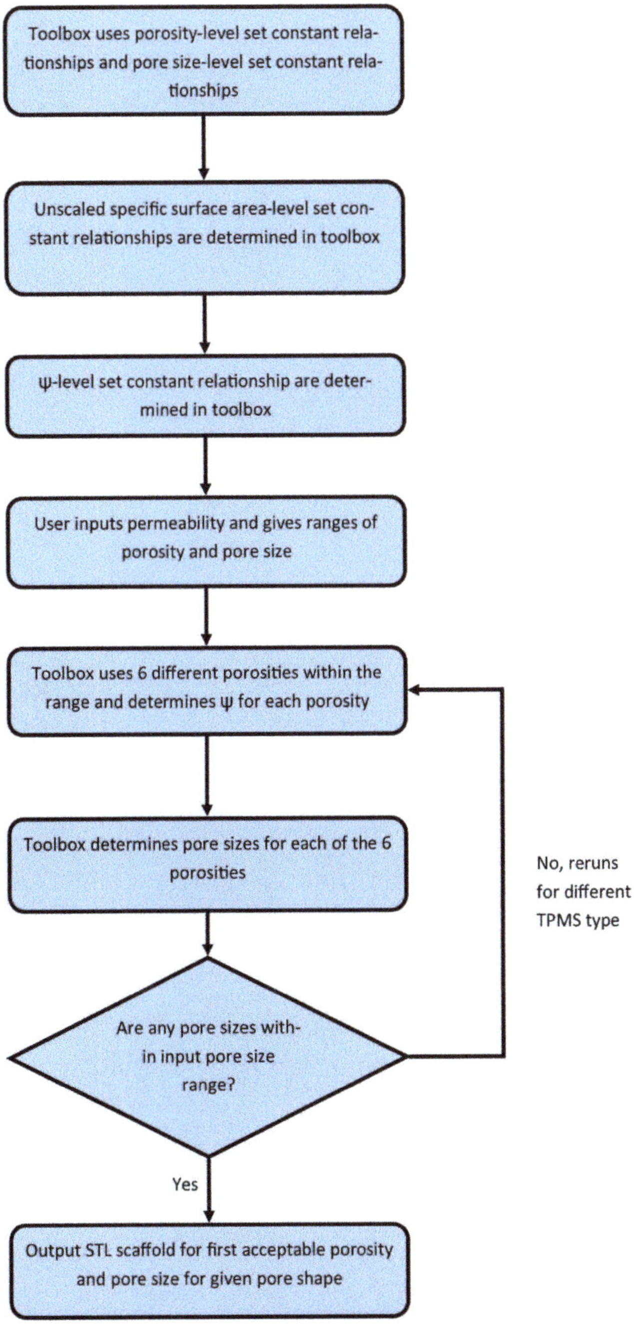

Figure 2. Flow chart for the process in which a scaffold is generated.

3. Results

3.1. Updated Kozeny–Carman Equation

Figure 3a relates the level-set constant to the unscaled specific surface area of the gyroid, where decreasing the level-set constant increases the unscaled specific surface area. Figure 3b shows the relationship between porosity and the level-set constant for the gyroid surface. Similarly, decreasing the level-set constant increases the porosity. This is the inverse function of Equation (2). Because both porosity and the unscaled specific surface area are a function of the level-set constant, if the scale factor and the Kozeny constant are known, then an estimation of permeability can be determined.

$$S(C,\alpha) = \frac{S_0(C)}{\alpha}, \tag{4}$$

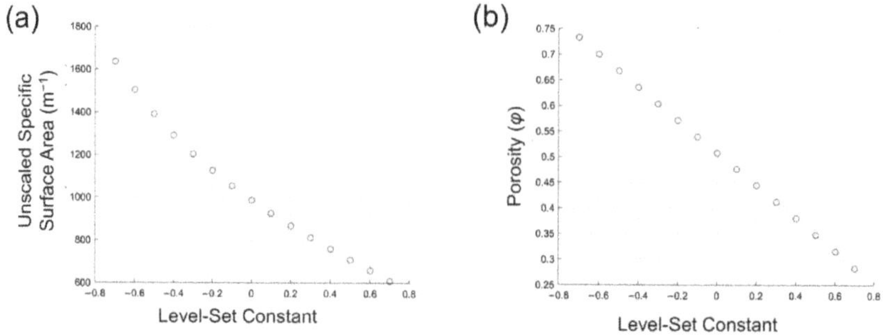

Figure 3. (**a**) Level-set constant vs. unscaled specific surface area for scaffolds, (**b**) level-set constant vs. porosity for scaffolds.

Porosity and ${d_p}^{2\pi}$ can be derived solely from the level-set constant, shown in Equations (2) and (3), and the unscaled specific surface area (S_0) can also be derived from the level-set constant, shown in Figure 3a. This carries that the majority of variables that make up the Kozeny–Carman equation (Equation (5)) can be derived from the level-set constant. Thus, a variable (ψ) of the Kozeny–Carman equation was designed. This variable (ψ) incorporates all of the variables solely derived from the level-set constant. The equation of this variable is shown in Equation (8), and the graphical representation of ψ is shown in Figure 4. Equation (5) shows the Kozeny–Carman equation, and Equation (6) shows the combination of Equations (4) and (5). Equation (7) shows the scale factor, α, which is used to correct the pore size.

$$\kappa = \beta \cdot \left(\frac{\varphi^3}{(1-\varphi)^2 \cdot S^2} \right), \tag{5}$$

$$\kappa = \beta \cdot \left(\frac{\varphi^3}{(1-\varphi)^2 \cdot {S_0}^2} \right) \cdot \alpha^2, \tag{6}$$

$$\alpha = \frac{d_p}{{d_p}^{2\pi} \cdot 2\pi}, \tag{7}$$

$$\begin{cases} \kappa = \beta \cdot \psi \cdot {d_p}^2 \\ \psi(C) = \frac{\varphi(C)^3 \cdot \left[{d_p}^{2\pi}(C) \cdot 2\pi\right]^2}{[1-\varphi(C)]^2 \cdot S_0(C)^2}, \end{cases} \tag{8}$$

$$d_p = \sqrt{\frac{\kappa}{\beta \cdot \psi}} \tag{9}$$

where ψ is the derived variable, β is the Kozeny constant, κ is the permeability, φ is the porosity, S_0 is the unscaled specific surface area, S is the specific surface area, α is the scale factor, ${d_p}^{2\pi}$ is the pore size of a structure prior to scaling, and d_p is the specified pore size.

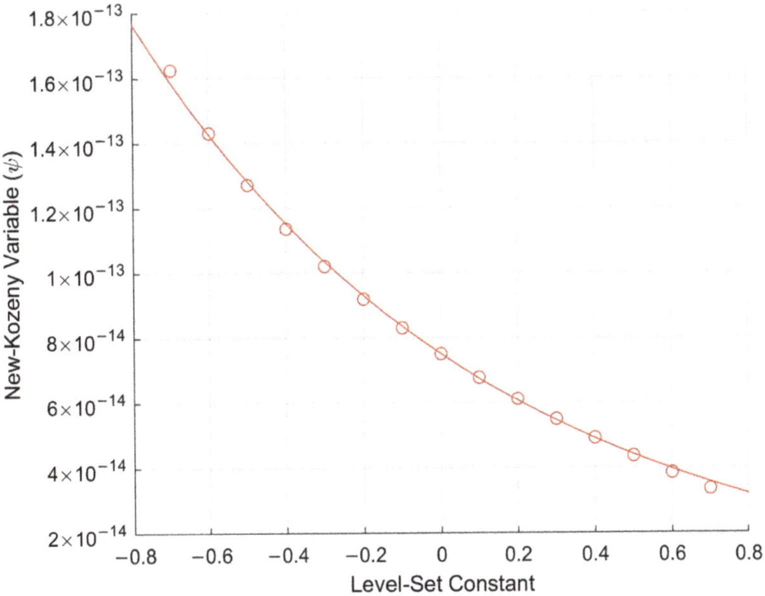

Figure 4. Relationship between level-set constant and ψ for gyroid structure by curve fitting.

The results of CFD simulations are shown in Table 1, where the average Kozeny constant over the 16 variations for each TPMS architecture is given.

Table 1. Average Kozeny constant for each TPMS type.

TPMS Type	Gyroid	Diamond	FKS	FRD	Primitive	Neovius
β	0.12 ± 0.021	0.13 ± 0.022	0.12 ± 0.019	0.17 ± 0.019	0.15 ± 0.032	0.12 ± 0.032

Table 1 indicates that all six TPMS types have β values close to 0.12. It is noteworthy that, although the primitive and Neovius structures have close geometric and visual similarities, they have different values for the Kozeny constant. This contrasts with the gyroid and FKS architectures, which are geometrically similar and also have the same β value. The typical value of β in the Kozeny–Carman equation was stated to have a value around ~0.2 [13]. The simulation results indicate that all the TPMS structures have significantly lower β values than 0.2.

3.2. MATLAB Toolbox for Scaffold Design

Based on the results of the porous geometry-influenced permeability above, a toolbox that can produce TPMS scaffolds based on a specified permeability have been developed in MATLAB R2023b (MathWorks, Netick, CA, USA), e.g., Figure 5 shows the user interface of the toolbox. The user will be required to enter the overall dimensions of the scaffold, the ranges for porosity and pore size, and the desired permeability. The ranges of porosity and pore size are important to enter, as the pore size and porosity of TPMS structures are highly reliant on appropriate manufacturing methods. The 'export' button will generate an STL file of the scaffold structure, which can then be used in 3D printing software, and the 'export inverse' button will generate an STL file of the geometric inverse, which can then be imported into a simulation software. The function for writing STL files was downloaded from the MATLAB Online file exchange [22].

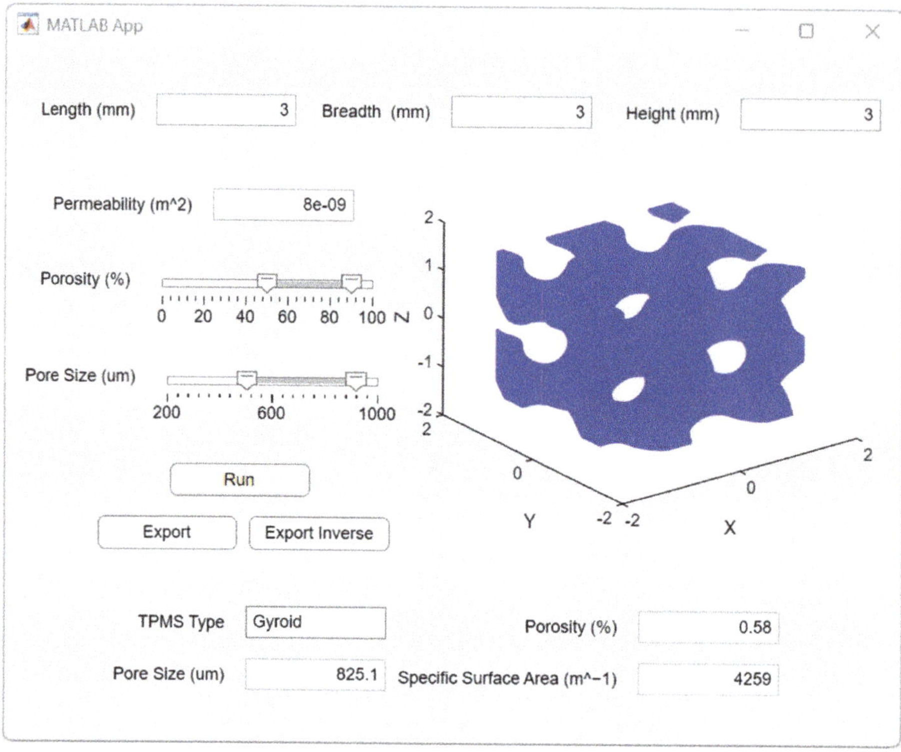

Figure 5. Screenshot of toolbox where a gyroid scaffold has been generated using a permeability input of 8×10^{-9} m^2 and geometric inputs of 3 mm in each direction.

4. Discussion

4.1. Permeability Comparison of TPMS Architectures

The values of β (Table 1) for each TPMS architecture were inputted into the toolbox, and this enabled the calculation of accurate permeabilities. Figure 6 shows the ranges of calculated permeabilities for both the lowest permeability for each TPMS architecture (pore size 400 μm and porosity 50%) and the highest permeability (pore size 700 μm and porosity 80%). From Figure 6, it can be seen that each of the scaffolds produces a higher permeability than cancellous bone [23]. This is expected, since TPMS structures have high permeabilities due to their low constriction factor, which, in turn, is due to their repeating units and their comparatively lower tortuosity. Another interesting observation from Figure 6 is the length of the ranges of the generatable permeabilities of the TPMS scaffolds. The F-RD, Neovius, and primitive scaffolds all have broader permeability ranges than the gyroid, diamond, and FKS scaffolds. The F-RD architecture also seems to have a significantly larger range of generatable permeabilities, allowing for the highest permeability to be produced. Additionally, it would seem that the Neovius structure is unique among the six scaffolds because of its lower range of generatable permeabilities, allowing for structures to be generated with permeabilities that are closer to that of cancellous bone. Because homogenous TPMS structures have lower constriction factors (leading to higher permeabilities), it is hypothesized that heterogenous-gradient TPMS structures could be used to achieve scaffold permeabilities even closer to that of cancellous bone.

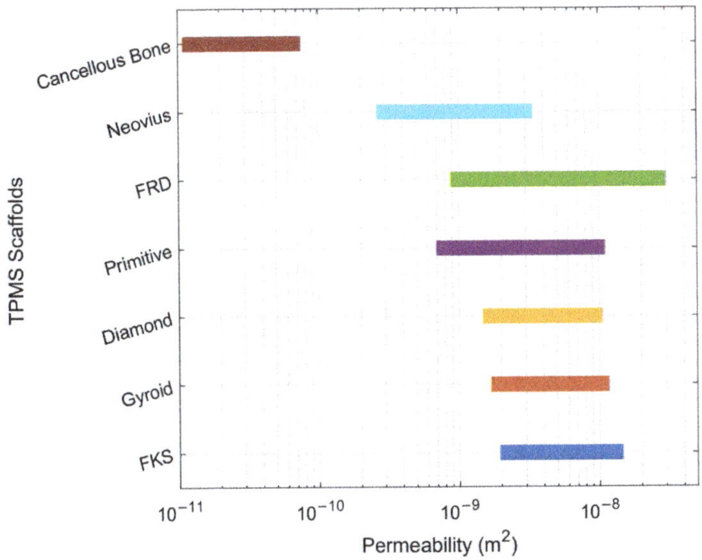

Figure 6. Range of permeabilities for TPMS scaffolds with comparative cancellous bone.

4.2. Limitations and Future Work

The simulation results indicate that the Kozeny–Carman equation, with the Kozeny constant (β) included, is an accurate way to estimate the permeability. This work has also shown that this Kozeny–Carman equation can be derived further into Equation (9), where the permeability can be determined through pore size, the Kozeny constant, and the new variable described in this paper. However, one major limitation of this method is that β was not constant for each TPMS architecture; decreasing pore size caused higher values of β. The variation in the value of β was negligible for five of the six TPMS architectures; however, using the same value of β for the F-RD architecture may cause slight inaccuracies in the calculated permeability. This work has also determined the relationship between specific surface area and the level-set constant, which was previously unknown, furthering the work of Walker et al., 2017.

Since the process for determining this new variable is automated, this constant could be determined for other common TPMS structures such as the lidinoid, CLP, or IWP. These structures could in future be implemented into the toolbox, if the range of attainable permeabilities needs to be increased. A further way to attain a larger range of permeabilities could be to include sheet-based lattices of the already available TPMS structures within the toolbox. This would first require new constant values and new level-set relationships to be determined. As described previously, heterogenous gradient TPMS structures could be used to achieve permeabilities closer to cancellous bone; both porosity and pore size gradients result in higher constriction factors and so should be tested in future. Additionally, combination TPMS structures would also result in a higher constriction factor and thus a lower permeability.

5. Conclusions

In this study, the Kozeny–Carman equation, which is used for calculating the permeability of porous material, has been updated for calculating the permeability of tissue engineering scaffolds with TPMS architectures. Based on this, a design toolbox has been developed which allows for the efficient generation of TPMS scaffolds with a tailored permeability. This will allow for tissue engineers to design and manufacture tissue engineering scaffolds with precisely controlled permeability.

Supplementary Materials: The following supporting information can be downloaded at: https://www.mdpi.com/article/10.3390/jmmp9030072/s1.

Author Contributions: Conceptualization, M.B.-T. and F.Z.; methodology, M.B.-T.; software, M.B.-T.; validation, M.B.-T. and F.Z.; formal analysis, M.B.-T.; investigation, M.B.-T.; data curation, M.B.-T.; writing—original draft preparation, M.B.-T.; writing—review and editing, F.Z., B.S., and P.N.; visualization, M.B.-T.; supervision, F.Z.; project administration, F.Z.; funding acquisition, F.Z. All authors have read and agreed to the published version of the manuscript.

Funding: This study was supported by EPSRC–Doctoral Training Partnership (DTP) scholarship (reference code: EP/T517987/1-2573181), and the Royal Society research grant (reference code: RGS\R2\212280).

Data Availability Statement: The data that support the findings of this study are available from the corresponding author upon reasonable request.

Conflicts of Interest: The authors declare no conflicts of interest.

References

1. Zhao, F.; van Rietbergen, B.; Ito, K.; Hofmann, S. Flow rates in perfusion bioreactors to maximise mineralisation in bone tissue engineering In Vitro. *J. Biomech.* **2018**, *79*, 232–237. [CrossRef] [PubMed]
2. Frantz, C.; Stewart, K.M.; Weaver, V.M. The extracellular matrix at a glance. *J. Cell Sci.* **2010**, *123*, 4195–4200. [CrossRef] [PubMed]
3. Dong, Z.; Zhao, X. Application of TPMS structure in bone regeneration. *Eng. Regen.* **2021**, *2*, 154–162. [CrossRef]
4. Sokollu, B.; Gulcan, O.; Konukseven, E.I. Mechanical Properties Comparison of Strut-Based and Triply Periodic Minimal Surface Lattice Structures Produced by Electron Beam Melting. *Addit. Manuf.* **2022**, *60*, 103199.
5. Lehder, E.F.; Ashcroft, I.A.; Wildman, R.D.; Ruiz-Cantu, L.A.; Maskery, I. A multiscale optimisation method for bone growth scaffolds based on triply periodic minimal surfaces. *Biomech. Model. Mechanobiol.* **2021**, *20*, 2085–2096. [CrossRef] [PubMed]
6. Castro, A.P.G.; Pires, T.; Santos, J.E.; Gouveia, B.P.; Fernandes, P.R. Permeability versus design in TPMS scaffolds. *Materials* **2019**, *12*, 1313. [CrossRef] [PubMed]
7. Ochoa, I.; Sanz-Herrera, J.A.; García-Aznar, J.M.; Doblaré, M.; Yunos, D.M.; Boccaccini, A.R. Permeability evaluation of 45S5 Bioglass®-based scaffolds for bone tissue engineering. *J. Biomech.* **2009**, *42*, 257–260. [CrossRef] [PubMed]
8. Zhao, F.; Xiong, Y.; Ito, K.; van Rietbergen, B.; Hofmann, S. Porous Geometry Guided Micro-mechanical Environment Within Scaffolds for Cell Mechanobiology Study in Bone Tissue Engineering. *Front. Bioeng. Biotechnol.* **2021**, *9*, 736489. [CrossRef] [PubMed]
9. Berg, C.F. Permeability Description by Characteristic Length, Tortuosity, Constriction and Porosity. *Transp. Porous Media* **2015**, *107*, 713–726. [CrossRef]
10. Rabbani, A.; Jamshidi, S. Specific surface and porosity relationship for sandstones for prediction of permeability. *Int. J. Rock Mech. Min. Sci.* **2014**, *71*, 25–32. [CrossRef]
11. Mauri, A.; Bertola, M. Alvascience: A New Software Suite for the QSAR Workflow Applied to the Blood–Brain Barrier Permeability. *Int. J. Mol. Sci.* **2022**, *23*, 12882. [CrossRef] [PubMed]
12. Zhang, H.L.; Yu, H.; Yuan, X.H.; Xu, H.Y.; Micheal, M.; Zhang, J.N.; Shu, H.L.; Wang, G.C.; Wu, H.A. Permeability prediction of low-resolution porous media images using autoencoder-based convolutional neural network. *J. Pet. Sci. Eng.* **2022**, *208*, 109589. [CrossRef]
13. The Carman-Kozeny Equation. Available online: https://www.doitpoms.ac.uk/tlplib/powder/carman.php (accessed on 22 May 2024).
14. Blanquer, S.B.G.; Werner, M.; Hannula, M.; Sharifi, S.; Lajoinie, G.P.R.; Eglin, D.; Hyttinen, J.; Poot, A.A.; Grijpma, D.W. Surface curvature in triply-periodic minimal surface architectures as a distinct design parameter in preparing advanced tissue engineering scaffolds. *Biofabrication* **2017**, *9*, 025001. [CrossRef] [PubMed]
15. Knödler, P.; Dreissigacker, V. Fluid Dynamic Assessment and Development of Nusselt Correlations for Fischer Koch S Structures. *Energies* **2024**, *17*, 688. [CrossRef]
16. Abou-Ali, A.M.; Lee, D.W.; Abu Al-Rub, R.K. On the Effect of Lattice Topology on Mechanical Properties of SLS Additively Manufactured Sheet-, Ligament-, and Strut-Based Polymeric Metamaterials. *Polymers* **2022**, *14*, 4583. [CrossRef] [PubMed]

17. Abou-Ali, A.M.; Al-Ketan, O.; Lee, D.W.; Rowshan, R.; Abu Al-Rub, R.K. Mechanical behavior of polymeric selective laser sintered ligament and sheet based lattices of triply periodic minimal surface architectures. *Mater. Des.* **2020**, *196*, 109100. [CrossRef]
18. Zou, S.; Mu, Y.; Pan, B.; Li, G.; Shao, L.; Du, J.; Jin, Y. Mechanical and biological properties of enhanced porous scaffolds based on triply periodic minimal surfaces. *Mater. Des.* **2022**, *219*, 110803. [CrossRef]
19. Lu, C.; Zhang, Y.; Aziz, M.; Wen, P.; Zhang, C.; Shen, Q.; Chen, F. Mechanical behaviors of multidimensional gradient gyroid structures under static and dynamic loading: A numerical and experimental study. *Addit. Manuf.* **2022**, *59*, 103187. [CrossRef]
20. Walker, J.M.; Bodamer, E.; Kleinfehn, A.; Luo, Y.; Becker, M.; Dean, D. Design and mechanical characterization of solid and highly porous 3D printed poly(propylene fumarate) scaffolds. *Prog. Addit. Manuf.* **2017**, *2*, 99–108. [CrossRef]
21. Ansys CFX-Pre User's Guide. 2021. Available online: https://dl.cfdexperts.net/cfd_resources/Ansys_Documentation/CFX/Ansys_CFX-Solver_Theory_Guide.pdf (accessed on 17 February 2025).
22. Sven. Stlwrite-Write ASCII or Binary STL Files. *MATLAB Central File Exchange.* 2025. Available online: https://www.mathworks.com/matlabcentral/fileexchange/20922-stlwrite-write-ascii-or-binary-stl-files (accessed on 11 January 2025).
23. Prakoso, A.T.; Basri, H.; Adanta, D.; Yani, I.; Ammarullah, M.I.; Akbar, I.; Ghazali, F.A.; Syahrom, A.; Kamarul, T. The Effect of Tortuosity on Permeability of Porous Scaffold. *Biomedicines* **2023**, *11*, 427. [CrossRef] [PubMed]

Journal of
Manufacturing and Materials Processing

MDPI

Article

Effect of Geometry on the Dissolution Behaviour of Complex Additively Manufactured Tablets

Seyedebrahim Afkhami [1], Meisam Abdi [1,*] and Reza Baserinia [2,*]

[1] School of Engineering and Sustainable Development, Faculty of Computing, Engineering, and Media, De Montfort University, Leicester LE1 9BH, UK; seyedebrahim.afkhami@dmu.ac.uk
[2] School of Engineering, College of Science and Engineering, University of Leicester, Leicester LE1 7RH, UK
* Correspondence: meisam.abdi@dmu.ac.uk (M.A.); rb662@leicester.ac.uk (R.B.)

Abstract: Additive manufacturing (AM) processes, such as fused deposition modelling (FDM), have emerged as transformative technologies in pharmaceutical manufacturing, enabling the production of drug delivery systems with complex and customised geometries. These advancements provide precise control over drug release profiles and facilitate the development of patient-specific medicines. This study investigates the dissolution behaviour of AM-fabricated tablets made from polyvinyl alcohol (PVA), a hydrophilic and biocompatible polymer widely used in drug delivery systems. The influence of the initial mass, surface area, and surface-area-to-volume ratio (S/V) on dissolution kinetics is evaluated for tablets with intricate geometries. Our findings demonstrate that these parameters, while critical for conventional tablet shapes, are insufficient to fully predict the dissolution behaviour of complex geometries. Furthermore, this study highlights how geometric modifications can enable the administration of the same drug dosage through sustained or immediate release profiles, offering enhanced versatility in drug delivery. By leveraging the geometric design freedom provided by AM technologies, this research underscores the potential for optimising drug delivery systems to improve therapeutic outcomes and patient compliance.

Keywords: additive manufacturing; tablet dissolution; fused deposition modelling; pharmaceutical 3D printing

1. Introduction

A large majority of the drugs developed are administered to patients in tablet form. To manufacture these tablets, the active pharmaceutical ingredient (API) is processed into powder and mixed with excipients (binders, gliders, bulking agents, etc.) to formulate the tablet ingredients [1,2]. The formulation is then introduced into dies of specific geometries before compaction using a pair of solid punches in a system known as a rotary tablet press [3–5]. Despite advancements in process control technologies used in these machines and better understanding of the parameters influencing the quality of tablet products [6,7], the process has an inherent limitation. Due to the nature of the compaction process, the geometry of the tablets that can be manufactured is limited to a few simple shapes such as square, round, oval, triangle, caplet, etc. Consequently, drug release and absorption are mostly controlled via processing parameters including compression stress and tablet size and the scope for changing the shape is limited. For instance, Molavi et al. [8] investigated the influence of size and mass on the dissolution of oblong, biconvex, round, and flat tablets and observed a higher rate of drug release for samples with a larger size and mass.

Comparing the shapes, they observed a higher dissolution efficiency for biconvex tablets compared to flat tablets.

Additive manufacturing (AM), also known as 3D printing, has evolved significantly over the past decade and its constituent technologies have been adopted by many industries, such as the aerospace, automotive, food, and fashion industries, for manufacturing prototypes and functional products with enhanced properties [9–11]. For pharmaceuticals, there is an increasing interest toward utilising AM for manufacturing tablets in both the R&D [12] and production stages [13]. The high degree of design freedom offered by AM allows for the development of complicated structures and multi-drug compositions, enhancing the effectiveness and safety of pharmaceuticals [9,14–16]. These technologies have been utilised by researchers to fabricate pharmaceutical solid dosage forms and evaluate the changes to the critical quality attributes of products. For example, Kyobula et al. [17] used the 3D inkjet printing method to fabricate fenofibrate in a beeswax carrier matrix of honeycomb architecture to control drug dissolution for personalised medicine applications. Crișan et al. [18] used different polyvinyl alcohol (PVA) particle sizes whilst preparing drug-loaded filaments and examined how it influenced the dissolution behaviour of tablets with open channels manufactured using the fused deposition modelling (FDM) technique. Moreover, McDonagh et al. [19] investigated the influence of infill geometry and density on the dissolution of tablets 3D printed using the Arburg Plastic Freeforming (APF) system and found the release profile to be affected significantly for different geometries with a similar infill density. While FDM is widely regarded as suitable for producing proof-of-concept prototypes, it has demonstrated substantial promise for the production of final pharmaceutical applications, particularly in personalised medicine [20].

AM provides an excellent route toward controlling the release profile and improving pharmaceutical efficacy through fabricating solids with complex geometries that cannot be manufactured alternatively. Such geometrical freedom facilitates optimising the dissolution behaviour of drugs [21]. There is a limited number of studies in the literature looking at geometrical effects on dissolution behaviour of tablets made by AM. Lee et al. [22] investigated the changes in the drug release rate of microplates of a similar mass but with different surface areas manufactured by piezoelectric inkjet printing and concluded that the release rate is predominantly influenced by the sample surface area. However, Zhang et al. [23] showed that dissolution behaviour is governed by the surface-area-to-volume ratio (S/V) of tablets. They examined the impact of porosity, pore shape, pore length, and pore alignment on drug release behaviour in 3D-printed constructs and recorded faster drug release rates for samples with larger porosities. They showed that under a fixed S/V the dissolution rate is not affected notably. Martinez et al. [24] used the stereolithographic (SLA) process to fabricate solid dosage forms in simple shapes, including cube, pyramid, and sphere. They also reported the S/V to be the main geometrical factor affecting the dissolution behaviour. In an earlier work, they ruled out the influence of surface area and stated that the S/V is the sole important parameter affecting the dissolution behaviour [25].

Polymer-based tablets often release drugs via complex mechanisms, including diffusion, swelling, erosion, or combinations thereof, depending on the material properties and environmental conditions. Diffusion-based models have gained significant attention due to their ability to predict drug release profiles effectively. These models range from analytical solutions of Fick's laws for simple geometries to advanced numerical models capable of addressing non-linear and dynamic systems. For example, Siepmann and Peppas [26] developed a model integrating Fickian diffusion and polymer swelling to describe drug release from hydrophilic matrices, which has since been applied widely. Similarly, Colombo et al. [27] utilised numerical simulations to capture time-dependent diffusion behaviour in swellable polymers. These approaches have proven efficient for understanding and predicting dissolution dynamics in 3D-printed tablets, offering insights into the effects of geometry, porosity, and material composition. Expanding on these methodologies could further enhance the precision of dissolution predictions, particularly for complex geometries fabricated using AM technologies.

The reviewed literature highlights the significance of understanding how tablet geometry influences dissolution behaviour, particularly in the context of 3D-printed polymer tablets. Despite advancements in AM technologies and prior research efforts, there remains a critical gap in the knowledge regarding the precise interplay between the tablet volume, surface area, and surface-to-volume ratio and the dissolution behaviour of 3D-printed tablets. Moreover, the current literature is mainly focused on simple geometries whilst additional factors arising from utilising complex shapes have not been considered. To address the gaps, our study aims to experimentally investigate the dissolution behaviour of FDM-printed cylindrical PVA tablets with varying geometrical features while maintaining a consistent mass, surface area, and surface-area-to-volume ratio. Through the comparison of concentration and dissolved mass profiles across time for each set of samples, our aim is to improve the critical impacts of geometry on tablet dissolution.

This holistic approach seeks to offer valuable insights into how tablet geometry influences the dynamics of drug release, thereby aiding in the refinement of geometric parameters and optimising tablet design for enhanced dissolution behaviour. While this study focuses on the dissolution behaviour of PVA tablets with different geometries without incorporating active pharmaceutical ingredients (APIs), the primary aim is to examine the fundamental effects of tablet geometry on dissolution dynamics. By using PVA as a model material, the study isolates the geometric influence on the dissolution process, allowing for a controlled investigation that excludes the complexity of drug-solvent interactions. This approach serves as a foundation for future research, where drug-loaded systems can be explored, building on these insights to optimise the design of drug delivery systems using 3D-printed polymers.

2. Materials

Polyvinyl alcohol (PVA), a water-soluble polymer with excellent biocompatibility, was selected for this study to evaluate how the geometry of the dosage form influences its dissolution behaviour, with distilled water used as the solvent. PVA plays a crucial role in controlled drug release systems through mechanisms primarily driven by matrix swelling and diffusion. Its hydrophilic nature allows water absorption, leading to polymer swelling and the formation of hydrated layers that facilitate the diffusion of the active pharmaceutical ingredient. This dual mechanism provides versatility in tailoring drug release profiles for both immediate and sustained delivery systems, making PVA an ideal material for studying the influence of geometry on drug release, particularly in relation to surface-to-volume ratios in polymeric matrices [28,29]. Furthermore, PVA's filaments are commercially available and commonly used in FDM additive manufacturing due to their water solubility, high biocompatibility, and low toxicity, which further underscores its desirability for controlled drug delivery applications [30,31].

Here, a commercially available filament of PVA (manufactured by Ultimaker, the Netherlands) with a diameter of 2.85 mm was used to fabricate different geometries. The PVA filament properties provided by the manufacturer are summarised in Table 1.

Table 1. Material properties of the PVA filament provided by Ultimaker.

PVA Filament Specifications	
Diameter	2.85 ± 0.10 mm
Colour	PVA Natural
Melting temperature	163 °C
3D-printing temperature	190–210 °C
Specific gravity	1.23 gr/cm^3

3. Methodology

3.1. Design and Fabrication of the Specimens

For this research project, four cylindrical geometries were considered to represent the solid dosage forms. These were the Solid Cylinder (SC), a cylindrical shell named Hollow Cylinder 1 (HC1), a cylinder with a central opening named Hollow Cylinder 2 (HC2), and a cylindrical shell with four square openings equally spaced in the middle named Hollow Cylinder 3 (HC3). The selection of the four geometries was specifically aimed at achieving significant variation in the mass, surface area, and surface-area-to-volume ratio to enable a detailed investigation of their effects on dissolution behaviour. The dimensions of the geometries were also tailored to the resolution capabilities of the 3D printer to ensure precise fabrication of the samples with the desired parameter variations. These geometries are demonstrated in Figure 1.

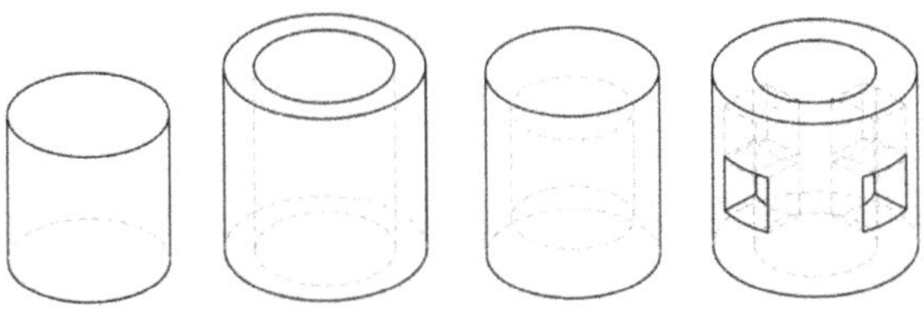

Figure 1. Example isometric projections of the four geometries. From left to right: Solid Cylinder (SC), Hollow Cylinder 1 (HC1), Hollow Cylinder 2 (HC2), and Hollow Cylinder 3 (HC3).

The design procedure was divided into three stages. First, geometries were manipulated to achieve a similar design mass of approximately 500 mg using the filament density (specific gravity) provided in Table 1. This was achieved by altering the dimensions of different designs and analysing the model in CAD software. Here, the modifications were such that the overall aspect ratios remained similar without considering the extreme cases where, for example, the geometry is elongated with a very small outer diameter. Similarly in the second stage, the dimensions were modified to achieve a similar surface area for the models (approximately 466 mm^2). In the third and final stage, the dimensions of the models were modified to achieve a similar S/V for the four geometries (approximately 0.75 mm^{-1}). It should be noted that the surface area considered here was the area of models that are initially exposed to the solvent when submerged, and in the case of HC3, it did not include the area of the hollow section inside the cylinder.

The four unique geometries corresponding to each set of samples were fabricated using an Ultimaker 3 FDM 3D printer with the layer height set at 0.1 mm and an infill density of 100% while utilising a 0.4 mm nozzle at a working temperature of 200 °C. This facilitated consistent layer bonding and ensured the structural integrity of the material. For each unique geometry, three replicate samples were 3D printed for further testing to improve the accuracy and reliability of the experimental analysis, making the total number of printed and analysed samples 36. An image of a set of geometries with similar surface areas is shown in Figure 2 as an example.

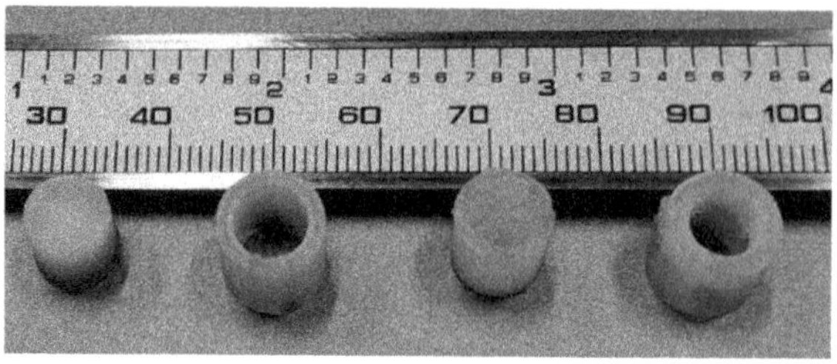

Figure 2. Example samples with similar surface areas. From left to right: Solid Cylinder (SC), Hollow Cylinder 1 (HC1), Hollow Cylinder 2 (HC2), and Hollow Cylinder 3 (HC3).

After production, each sample was weighted using a precision scale to measure its actual mass, which was averaged for each geometry. The design and measured properties of the samples fabricated for the three sets of designs are summarised in Table 2. In the table, the design values are rounded to the nearest integer and the tolerance values indicated for the averaged measured mass are the standard deviations. The mass error values listed in the final column compare the design values with the averaged measured masses. Discrepancies in mass can directly lead to discrepancies in concentration values, potentially affecting the accuracy of the dissolution profiles and the interpretation of results.

Table 2. The design and measured properties for manufactured specimens of four different shapes fabricated using FDM.

Geometry		Designed Mass (mg)	Designed Surface Area (mm^2)	Surface-Area-to-Volume (mm^{-1})	Average Measured Mass (mg)	Mass Error (%)
Similar Mass (M)	Solid Cylinder (SC-M)	502	302	0.75	487 ± 2.9	2.9
	Hollow Cylinder 1 (HC1-M)	501	614	1.53	500 ± 11.9	0.1
	Hollow Cylinder 2 (HC2-M)	497	380	0.95	500 ± 6.5	0.5
	Hollow Cylinder 3 (HC3-M)	504	605	1.50	503 ± 5.8	0.1
Similar Surface Area (S)	Solid Cylinder (SC-S)	950	466	0.61	901 ± 10.5	5.1
	Hollow Cylinder 1 (HC1-S)	371	466	1.57	359 ± 2.6	3.2
	Hollow Cylinder 2 (HC2-S)	758	466	0.77	732 ± 8.5	3.4
	Hollow Cylinder 3 (HC3-S)	346	468	1.69	324 ± 10.9	6.2
Similar Surface-Area-to-Volume Ratio (S/V)	Solid Cylinder (SC-SV)	497	300	0.75	470 ± 4	5.4
	Hollow Cylinder 1 (HC1-SV)	958	528	0.75	922 ± 12.3	3.8
	Hollow Cylinder 2 (HC2-SV)	981	604	0.77	910 ± 19.6	7.2
	Hollow Cylinder 3 (HC3-SV)	1489	923	0.77	1404 ± 35.1	5.6

For simplification in the following sections, the samples with similar mass are labelled with the letter M, the samples with a similar surface area with S, and the samples with a similar surface-area-to-volume ratio with SV.

3.2. Dissolution Testing

Following the fabrication of the samples using the 3D printer, their dissolution behaviour was examined with a PTWS 120 USP (IV) dissolution tester manufactured by Pharma Test, Germany. Each sample was placed in a 250 mL distilled water bath as the solvent and kept at a constant temperature of 37 °C to mimic the physiological conditions of the body. During testing, the solutions were stirred at a constant speed of 50 rpm to ensure consistent mixing over the testing period of up to 8 h. At different intervals, 3 mL samples were extracted from the solutions and filtered with a Sartorius CA 0.45 filter. The absorbance of the dissolved PVA was then determined using an Evolution 60S UV-visible spectrophotometer manufactured by Thermo Scientific, US. Following the measurements, the calibration curve presented in Figure 3a was used to convert the absorbance values to the corresponding concentrations in the solution.

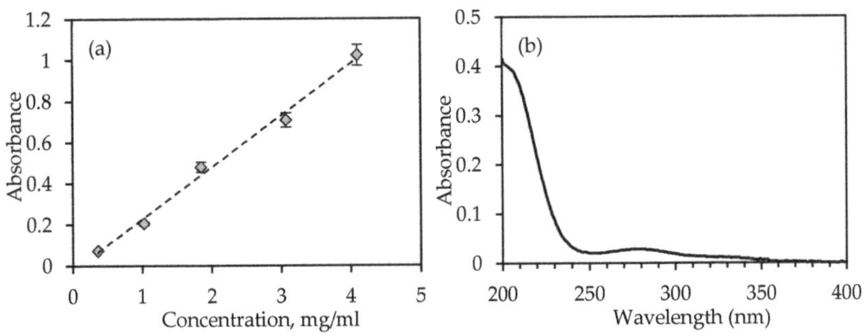

Figure 3. (**a**) The calibration curve for converting the absorbance to the concentration of PVA dissolved in water (coefficient of determination, $R^2 = 0.99$), and (**b**) the UV absorption standard curve for the PVA.

Prior to testing, the peak absorbance of the 3D-printed PVA was measured with the UV-visible spectrophotometer, as shown in Figure 3b. The peak absorbance was found to be at the UV wavelength of 206 nm followed by a minor peak at 278 nm. UV-visible spectrometry experiments were conducted at peak wavelengths of 206 where the material absorbance is maximum.

4. Results and Discussion

Following the dissolution testing of the 3D-printed geometries, the changes in the average concentration and dissolved percentage of PVA in water over time were calculated for the three sets of samples. The results obtained are compared in the following subsections.

4.1. Dissolution Behaviour of Samples with a Similar Mass

The changes in the average concentration and average dissolved percentage of PVA samples with a similar initial mass are shown in Figure 4, with the error bars indicating the standard deviation of three repeat measurements. Logically, the two figures are very similar and minor discrepancies are the result of a small difference in the initial mass of the samples, which influenced the dissolved percentage values.

The results show a similar release profile for HC1-M and HC3-M samples whereby 90% of the polymers were dissolved in under 200 min, whilst for SC-M and HC2-M the time required to reach the same threshold was double. The results highlight that despite having a similar overall mass or volume, the geometry of the tablets can be manipulated to achieve very different release profiles. This suggests that the same overall drug concentration can be administered as immediate release or controlled release by carefully tailoring the geometry of the solid dosage form.

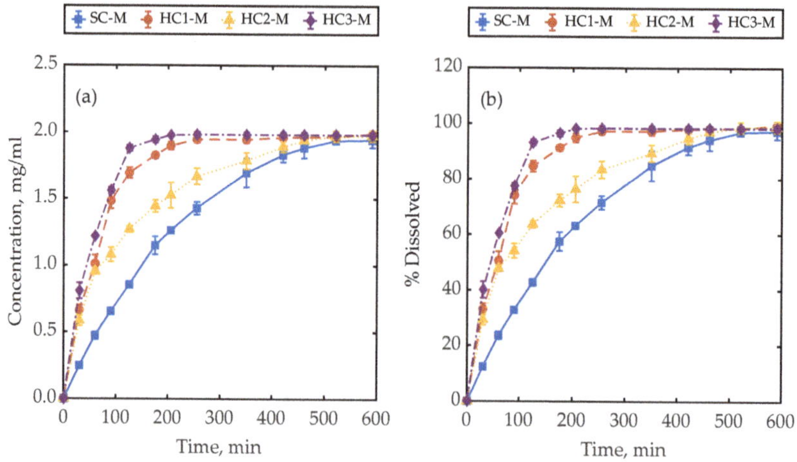

Figure 4. The changes in (**a**) average concentration and (**b**) percentage dissolved over time for samples with a similar mass. The error bars are generated using the standard deviation of the three repeat measurements.

As discussed earlier, the surface area and surface-area-to-volume ratio are widely recognised by researchers as critical factors influencing the dissolution of tablets. These factors have been examined for samples with a similar mass. For HC1-M and HC3-M, the values of these parameters are comparable and significantly larger than those of the other geometries (Table 2). The similarities in the dissolution behaviour of these two samples (Figure 4) align well with findings reported in the literature. Furthermore, Figure 4 shows that the SC-M sample, which has the lowest surface area within the set of four samples of a similar mass, exhibits the lowest diffusion rate, which is again consistent with the conclusions drawn by other researchers in previous studies [22,32,33].

The correlations between the surface area, the S/V, and the dissolution behaviour are further investigated by examining the release profiles of samples with a similar surface area and similar S/V in the following subsections.

4.2. Dissolution Behaviour of Samples with a Similar Surface Area

The changes in the concentration of the dissolved polymer and the polymer percentage dissolved over time for samples with a similar surface area are shown in Figure 5. Unlike in the previous set of samples, where the concentration–time graph closely resembled the % dissolved–time graph due to similar sample masses, in this set HC1-S and HC3-S have significantly smaller masses compared to SC-S and HC2-S. As a result, these samples with a smaller mass contribute less to the concentration of the solvent, which is reflected in the lower curves on the concentration–time graph. However, their smaller mass allows them to dissolve more rapidly percentagewise, resulting in higher curves on the % dissolved–time graph. Similar to previous results, for the HC1-S and HC3-S geometries, the release profiles suggest a rapid release response, whilst for SC-S and HC2-S the time required to fully dissolve the samples was larger and polymer release was sustained over a longer period (Figure 5b).

Comparing all four profiles shows that, despite having similar surface areas, the dissolution behaviours of the samples are significantly different. However, if these results are compared with respect to the surface-area-to-volume ratios, for samples with a larger

S/V the time required to reach a fixed percentage released is shorter. This is consistent with the results obtained by Martinez et al. [24]. However, the varying masses and overall sizes of the samples examined in this section preclude a reliable comparison based solely on the percentage dissolved. Therefore, the apparent correlations between the dissolution behaviour and S/V may be inaccurate.

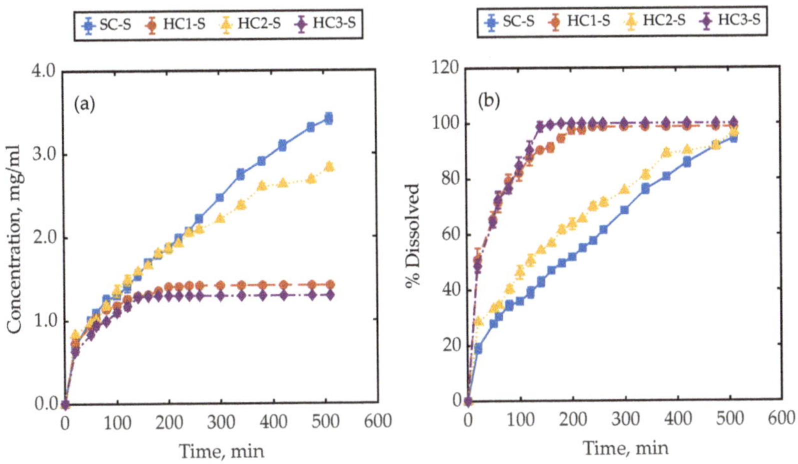

Figure 5. The changes in (**a**) average concentration and (**b**) percentage dissolved over time for samples with a similar surface area. The error bars are generated using the standard deviation of the three repeat measurements.

To evaluate the correlations, the results were compared with respect to changes in the concentration of the polymer in the solution, as shown in Figure 5a. Despite observing a sustained release behaviour for SC-S and HC2-S relative to others (Figure 5b), larger average concentrations were recorded for these geometries over the measurement period. In late stages of the dissolution measurement, this is mostly due to larger overall masses of the samples, as indicated in Table 2. However, in early stages, the higher concentration values measured for SC-S and HC2-S suggest dependency on factors other than the initial surface area or the initial S/V individually. These samples have a similar initial surface area but a smaller S/V. According to conclusions made by researchers previously, these samples should exhibit the same behaviour if the surface area is considered or reduced dissolution if the S/V is considered. Yet the results contradict both cases. Therefore, once again, it cannot be stated that the dissolution behaviour is solely a function of the initial surface area or initial S/V.

These results further consolidate the idea of achieving a desired release mechanic through geometry alteration rather than formulation or process parameter changes. Additionally, it is highlighted that the common approach of reporting the release profiles in terms of percentage release can present misleading information about dissolution kinetics.

4.3. Dissolution Behaviour of Samples with a Similar Surface-Area-to-Volume Ratio

The changes in the concentration and percentage of polymer dissolved over time for samples with a similar S/V are shown in Figure 6. From Figure 6a, it can be observed that the concentration changes differ among the four samples with a similar S/V, which can be attributed to the variations in their mass and surface area. The HC3-SV specimen, having a significantly larger mass than the other specimens, results in a much higher concentration level. The concentration–time trends also align with the Fickian diffusion model, where the diffusion rate is proportional to the surface area of the specimens. This is evident from the slopes of the concentration–time curves at the start of dissolution: specimens with larger surface areas (e.g., HC3-SV) exhibit steeper initial slopes (i.e., higher diffusion rates), while those with smaller surface areas (e.g., SC-SV) demonstrate gentler slopes (i.e., lower diffusion rates), highlighting the influence of surface area on dissolution.

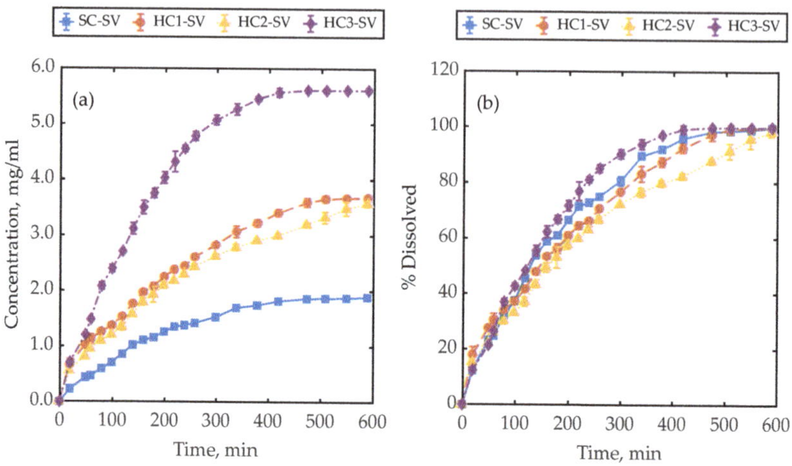

Figure 6. The changes in (**a**) average concentration and (**b**) percentage dissolved over time for samples with a similar surface-area-to-volume ratio. The error bars are generated using the standard deviation of the three repeat measurements.

Although the general trend in the concentration–time graph is a gradual reduction in slope as concentration increases due to the decreasing surface area of the specimens during dissolution, an exception is noted for the HC2-SV specimen. Around 380 min after the start of dissolution, HC2-SV shows an unexpected increase in the slope of the concentration–time curve. This is likely due to a sudden increase in the effective surface area as the internal hole of this cylindrical specimen, initially inaccessible to the solvent, becomes exposed. A similar trend—an abrupt increase in the slope of the concentration–time curve—can also be observed for the HC2-S specimen in Figure 5a, suggesting the same underlying cause. Once more, this underscores the importance of geometry design and the interplay between different parameters such as the mass, surface area, and topology of the specimens in determining dissolution kinetics.

It can be seen from Figure 6b that, despite having very different mass and surface area values, this set of specimens demonstrates a closer trend in terms of percentage dissolved over time compared with the previous sets. This observation aligns with findings by other researchers [24], suggesting that dissolution behaviour (in terms of percentage dissolved) is primarily influenced by the S/V, and the difference observed could be due to variation in the accessibility of the solvent to the internal volume of the samples. A slightly different trend is noted for the HC2-SV specimen, which is attributed to its specific design featuring an internal hole that is initially not exposed to the solvent. This unique topology impacts the dissolution behaviour by delaying solvent access to certain regions of the sample, demonstrating the interplay between geometric features and dissolution kinetics.

4.4. Further Remarks and Discussion

The dissolution behaviour of pharmaceutical tablets is a multifaceted process influenced by geometric parameters, surface properties, and solvent interactions. While earlier sections have explored specific geometric influences such as the surface area and S/V, this section synthesises these findings to provide a broader understanding of their combined impact. Figure 7 compares the average time to 50% dissolution (half dissolution time) for all 12 unique geometries, enabling a ranking of the geometries based on their dissolution performance. It can be observed that the SC-S specimen exhibits the longest half dissolution time, which correlates with its having the lowest S/V (0.61 mm^{-1}) among the 12 samples, as shown in Table 2. Conversely, the two samples with the shortest half dissolution times, HC3-S and HC1-S, have the highest S/V, 1.69 mm^{-1} and 1.57 mm^{-1}, respectively. For samples with a similar S/V, the half dissolution times remain within the same range, reinforcing the critical role of the S/V in determining dissolution rates. These findings underscore how surface area, mass, and topology can be tailored to achieve a desired dissolution response, offering valuable insights for designing solid dosage forms with specific dissolution characteristics.

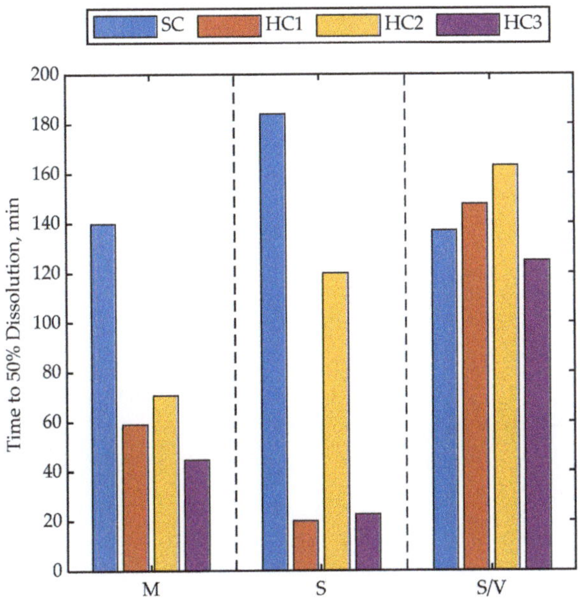

Figure 7. Time to 50% dissolution for all studied specimens.

To further substantiate the dissolution behaviour observed in this study, the data for the samples with a similar S/V were fitted to the Higuchi model, a well-established mathematical framework describing drug release from polymeric matrices governed by Fickian diffusion [26,34]. The Higuchi model expresses the cumulative release as proportional to the square root of time:

$$M_t = k_H t^{1/2} \tag{1}$$

where M_t is the amount of drug released at time t (% dissolved), and k_H is the Higuchi dissolution constant. This model assumes constant diffusivity and a homogeneous polymer matrix [35]. As shown in Figure 8, the linearity of the plotted curves for cumulative drug release (M_t) versus the square root of time ($t^{1/2}$) for all four unique specimens confirms their conformity to the Higuchi model, demonstrating that the dissolution mechanism for PVA tablets is primarily diffusion-controlled. The slope, intersect, and coefficient of determination (R^2) for the lines fitted to the data using linear regression (LR) are summarised in Table 3. While the similarity in the slopes of these lines (Higuchi dissolution constants) indicates comparable dissolution behaviour for specimens with equivalent S/V ratios, discrepancies in the magnitude of the slopes suggest that additional factors, such as initial mass, material distribution, and internal topology, also play a significant role in influencing dissolution. These findings align with prior research indicating that the Fickian diffusion mechanism dominates dissolution in hydrophilic polymer matrices like PVA [27,36], reinforcing the suitability of this model for describing the system studied.

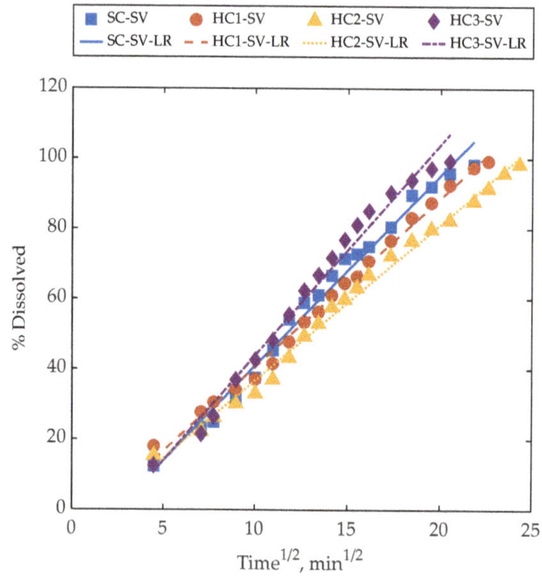

Figure 8. Linear regression (LR) plots of cumulative drug release (% dissolved) versus the square root of time for specimens with a similar S/V fitted to the Higuchi model.

Table 3. The slope, intercept, and coefficient of determination, R2, for the lines fitted to the data using linear regression.

Sample	Slope	Intercept	R^2
SC-SV	5.42	−13.16	0.99
HC1-SV	4.81	−7.41	0.99
HC2-SV	4.45	−7.70	0.99
HC3-SV	6.01	−16.15	0.98

According to Fick's first law of diffusion [37], the diffusion rate (i.e., the rate of change in concentration over time, represented by the slope of the concentration–time graph) is directly proportional to the surface area when comparing the dissolution of specimens at a fixed concentration. In the case of the presented experiments, such comparisons can only be made in the very early stages of dissolution when the concentration is near zero and the surface area of the specimens has not significantly changed due to dissolution, typically within the first 10 min. Theoretically, during this interval, the slopes of the concentration–time graphs for all specimens with a similar surface area should be comparable. However, this effect is not captured in the presented experimental data due to the need for a much higher number of concentration measurements within a short period of time. Figure 5b shows that shortly after the start of dissolution, within the first 50 m, a significant percentage of the polymer samples dissolve: 25–30% for the SC-S and HC2-S samples, and approximately 60% for the HC1-S and HC-3S samples. This indicates that the surface area of the specimens changes significantly from the starting point. Therefore, considering the original surface area as the sole factor affecting dissolution behaviour is invalid as the change in the shape of the samples over time is not governed solely by the original surface area. The same holds true for changes in the S/V over the experimentation period (Figure 6b).

In this context, S/V values provide further insights into the dissolution behaviour of the specimens. Higher S/V values theoretically imply that a larger percentage of the specimen is exposed to the solvent, resulting in enhanced release behaviour, as observed for the HC1^{-}S and HC^{-}3S samples in Figure 5b. For the same reason, higher S/V values lead to a higher rate of change in geometry and a reduction in surface area, consequently decreasing the diffusion rate (slope of concentration–time curves in Figure 5a). This effect is visible in Figure 5a, where the HC1^{-}S and HC^{-}3S samples, which have higher S/V values, experience a rapid change in the diffusion rate compared to the other two specimens with lower S/V values and a more sustained release behaviour over time. Consistently, for samples with a similar S/V (Figure 6a), the initial change in concentration is similar, supporting this argument; however, as the dissolution continues, the evolution of the shape of the samples and, thus, changes in the S/V become dissimilar, resulting in different concentrations.

5. Summary and Conclusions

Additive manufacturing (AM) has the potential to transform the design and production of pharmaceutical tablets by enabling the fabrication of complex geometries. These advancements offer opportunities to tailor drug release profiles, addressing the limitations of conventional manufacturing methods. This study explored the dissolution behaviour of 3D-printed polyvinyl alcohol (PVA) tablets, which were fabricated using fused deposition modelling (FDM), focusing on the effects of the geometry, surface area, and surface-area-to-volume ratio (S/V).

The findings demonstrate that while the surface area and S/V are critical factors for simple geometries, they are insufficient to fully predict the dissolution behaviour of complex shapes. Complex geometries exhibited unique dissolution dynamics influenced by solvent accessibility and internal topology, challenging the conventional reliance on the surface area or S/V as sole predictors. This highlights the importance of considering additional parameters when designing advanced drug delivery systems.

A key outcome of this research is the potential for geometric tailoring to provide precise control over drug release profiles. Tablets with an identical mass but varying geometries demonstrated distinct dissolution behaviours, enabling the delivery of the same drug concentration through either immediate or sustained release by altering the design. This approach offers a promising pathway for optimising drug release without modifying formulation or processing conditions.

The implications of this work extend beyond simple geometry. For diffusion-controlled systems like PVA-based tablets, higher S/V ratios enhance solvent penetration and diffusion, while erosion- and swelling-controlled mechanisms benefit from similar geometric optimisations. These findings underline the critical role of geometry in achieving desired dissolution outcomes and pave the way for integrating complex shapes into personalised and multi-drug delivery systems.

Future research should focus on incorporating active pharmaceutical ingredients (APIs) to evaluate how geometry influences active drug release mechanisms. Additionally, exploring multi-drug tablet designs will help expand the application of AM in complex therapeutic scenarios. Optimisation techniques, such as machine learning and evolutionary algorithms, could further enhance the design of tablet geometries to achieve specific release profiles.

This study highlights the transformative potential of AM for pharmaceutical manufacturing. By leveraging geometric design freedom and integrating experimental and computational approaches, AM could enable advanced drug delivery systems tailored to individual therapeutic needs, enhancing both efficacy and patient compliance.

Supplementary Materials: The following supporting information can be downloaded at: https://www.mdpi.com/article/10.3390/jmmp9010011/s1, Dissolution Data are in the Supplementary Materials.

Author Contributions: Conceptualization, S.A., M.A. and R.B.; methodology, S.A.; validation, S.A.; formal analysis, S.A.; investigation, S.A.; resources, R.B.; writing—original draft preparation, S.A.; writing—review and editing, R.B. and M.A.; visualisation, S.A. and R.B.; supervision, R.B. and M.A.; project administration, R.B. and M.A.; funding acquisition, R.B. All authors have read and agreed to the published version of the manuscript.

Funding: This study was funded by the Faculty of Computing, Engineering and Media at De Montfort University.

Data Availability Statement: The original contributions presented in this study are included in the article/Supplementary Materials. Further inquiries can be directed to the corresponding authors.

Conflicts of Interest: The authors declare no conflicts of interest.

References

1. van der Merwe, J.; Steenekamp, J.; Steyn, D.; Hamman, J. The Role of Functional Excipients in Solid Oral Dosage Forms to Overcome Poor Drug Dissolution and Bioavailability. *Pharmaceutics* **2020**, *12*, 393. [CrossRef] [PubMed]
2. Jakubowska, E.; Ciepluch, N. Blend Segregation in Tablets Manufacturing and Its Effect on Drug Content Uniformity—A Review. *Pharmaceutics* **2021**, *13*, 1909. [CrossRef] [PubMed]
3. Wünsch, I.; Friesen, I.; Puckhaber, D.; Schlegel, T.; Finke, J.H. Scaling Tableting Processes from Compaction Simulator to Rotary Presses-Mind the Sub-Processes. *Pharmaceutics* **2020**, *12*, 310. [CrossRef]
4. Schomberg, A.K.; Kwade, A.; Finke, J.H. The Challenge of Die Filling in Rotary Presses—A Systematic Study of Material Properties and Process Parameters. *Pharmaceutics* **2020**, *12*, 248. [CrossRef]
5. Schomberg, A.K.; Kwade, A.; Finke, J.H. Modeling gravity filling of dies on a rotary tablet press. *Powder Technol.* **2023**, *413*, 117998. [CrossRef]
6. Sinka, I.; Motazedian, F.; Cocks, A.; Pitt, K. The effect of processing parameters on pharmaceutical tablet properties. *Powder Technol.* **2009**, *189*, 276–284. [CrossRef]
7. Manley, L.; Hilden, J.; Valero, P.; Kramer, T. Tablet Compression Force as a Process Analytical Technology (PAT): 100% Inspection and Control of Tablet Weight Uniformity. *J. Pharm. Sci.* **2019**, *108*, 485–493. [CrossRef]
8. Molavi, F.; Hamishehkar, H.; Nokhodchi, A. Impact of Tablet Shape on Drug Dissolution Rate Through Immediate Released Tablets. *Adv. Pharm. Bull.* **2020**, *10*, 656–661. [CrossRef]
9. Ngo, T.D.; Kashani, A.; Imbalzano, G.; Nguyen, K.T.Q.; Hui, D. Additive manufacturing (3D printing): A review of materials, methods, applications and challenges. *Compos. Part B Eng.* **2018**, *143*, 172–196. [CrossRef]
10. Prashar, G.; Vasudev, H.; Bhuddhi, D. Additive manufacturing: Expanding 3D printing horizon in industry 4.0. *Int. J. Interact. Des. Manuf. IJIDeM* **2023**, *17*, 2221–2235. [CrossRef]
11. Iftekar, S.F.; Aabid, A.; Amir, A.; Baig, M. Advancements and limitations in 3D printing materials and technologies: A critical review. *Polymers* **2023**, *15*, 2519. [CrossRef] [PubMed]
12. Milliken, R.L.; Quinten, T.; Andersen, S.K.; Lamprou, D.A. Application of 3D printing in early phase development of pharmaceutical solid dosage forms. *Int. J. Pharm.* **2024**, *653*, 123902. [CrossRef]
13. Li, C.; Pisignano, D.; Zhao, Y.; Xue, J. Advances in Medical Applications of Additive Manufacturing. *Engineering* **2020**, *6*, 1222–1231. [CrossRef]
14. Khaled, S.A.; Alexander, M.R.; Irvine, D.J.; Wildman, R.D.; Wallace, M.J.; Sharpe, S.; Yoo, J.; Roberts, C.J. Extrusion 3D Printing of Paracetamol Tablets from a Single Formulation with Tunable Release Profiles Through Control of Tablet Geometry. *AAPS Pharmscitech* **2018**, *19*, 3403–3413. [CrossRef]
15. Vora, L.K.; Gholap, A.D.; Jetha, K.; Thakur, R.R.S.; Solanki, H.K.; Chavda, V.P. Artificial intelligence in pharmaceutical technology and drug delivery design. *Pharmaceutics* **2023**, *15*, 1916. [CrossRef] [PubMed]
16. Li, Q.; Wen, H.; Jia, D.; Guan, X.; Pan, H.; Yang, Y.; Pan, W. Preparation and investigation of controlled-release glipizide novel oral device with three-dimensional printing. *Int. J. Pharm.* **2017**, *525*, 5–11. [CrossRef] [PubMed]

17. Kyobula, M.; Adedeji, A.; Alexander, M.R.; Saleh, E.; Wildman, R.; Ashcroft, I.; Gellert, P.R.; Roberts, C.J. 3D inkjet printing of tablets exploiting bespoke complex geometries for controlled and tuneable drug release. *J. Control. Release* **2017**, *261*, 207–215. [CrossRef]
18. Crișan, A.G.; Porfire, A.; Ambrus, R.; Katona, G.; Rus, L.M.; Porav, A.S.; Ilyés, K.; Tomuță, I. Polyvinyl Alcohol-Based 3D Printed Tablets: Novel Insight into the Influence of Polymer Particle Size on Filament Preparation and Drug Release Performance. *Pharmaceuticals* **2021**, *14*, 418. [CrossRef]
19. McDonagh, T.; Belton, P.; Qi, S. An investigation into the effects of geometric scaling and pore structure on drug dose and release of 3D printed solid dosage forms. *Eur. J. Pharm. Biopharm.* **2022**, *177*, 113–125. [CrossRef] [PubMed]
20. Cailleaux, S.; Sanchez-Ballester, N.M.; Gueche, Y.A.; Bataille, B.; Soulairol, I. Fused Deposition Modeling (FDM), the new asset for the production of tailored medicines. *J. Control. Release* **2021**, *330*, 821–841. [CrossRef]
21. Serajuddin, A.T.M. Challenges, current status and emerging strategies in the development of rapidly dissolving FDM 3D-printed tablets: An overview and commentary. *Admet Dmpk* **2023**, *11*, 33–55. [CrossRef]
22. Lee, B.K.; Yun, Y.H.; Choi, J.S.; Choi, Y.C.; Kim, J.D.; Cho, Y.W. Fabrication of drug-loaded polymer microparticles with arbitrary geometries using a piezoelectric inkjet printing system. *Int. J. Pharm.* **2012**, *427*, 305–310. [CrossRef] [PubMed]
23. Zhang, B.; Gleadall, A.; Belton, P.; Mcdonagh, T.; Bibb, R.; Qi, S. New insights into the effects of porosity, pore length, pore shape and pore alignment on drug release from extrusionbased additive manufactured pharmaceuticals. *Addit. Manuf.* **2021**, *46*, 102196. [CrossRef]
24. Martinez, P.R.; Goyanes, A.; Basit, A.W.; Gaisford, S. Influence of Geometry on the Drug Release Profiles of Stereolithographic (SLA) 3D-Printed Tablets. *AAPS Pharmscitech* **2018**, *19*, 3355–3361. [CrossRef]
25. Goyanes, A.; Martinez, P.R.; Buanz, A.; Basit, A.W.; Gaisford, S. Effect of geometry on drug release from 3D printed tablets. *Int. J. Pharm.* **2015**, *494*, 657–663. [CrossRef] [PubMed]
26. Siepmann, J.; Peppas, N.A. Modeling of drug release from delivery systems based on hydroxypropyl methylcellulose (HPMC). *Adv. Drug Deliv. Rev.* **2001**, *48*, 139–157. [CrossRef] [PubMed]
27. Colombo, P.; Bettini, R.; Santi, P.; De Ascentiis, A.; Peppas, N.A. Swellable matrices for controlled drug delivery: Gel-layer behavior, mechanisms and optimal performance. *Pharm. Sci. Technol. Today* **1996**, *1*, 202–208. [CrossRef]
28. Zahra, F.T.; Quick, Q.; Mu, R. Electrospun PVA fibers for drug delivery: A review. *Polymers* **2023**, *15*, 3837. [CrossRef]
29. Otarbayeva, S.; Berillo, D. Poly (Vinyl Alcohol) Drug and PVA–Drug–Surfactant Complex Organogel with Dimethyl Sulfoxide as a Drug Delivery System. *Gels* **2024**, *10*, 753. [CrossRef] [PubMed]
30. Basa, B.; Jakab, G.; Kállai-Szabó, N.; Borbás, B.; Fülöp, V.; Balogh, E.; Antal, I. Evaluation of biodegradable PVA-based 3D printed carriers during dissolution. *Materials* **2021**, *14*, 1350. [CrossRef]
31. Al-Taie, A.; Pan, J.; Polak, P.; Barer, M.R.; Han, X.; Abbott, A.P. Mechanical properties of 3-D printed polyvinyl alcohol matrix for detection of respiratory pathogens. *J. Mech. Behav. Biomed. Mater.* **2020**, *112*, 104066. [CrossRef] [PubMed]
32. Windolf, H.; Chamberlain, R.; Quodbach, J. Predicting drug release from 3D printed oral medicines based on the surface area to volume ratio of tablet geometry. *Pharmaceutics* **2021**, *13*, 1453. [CrossRef]
33. Tsunematsu, H.; Hifumi, H.; Kitamura, R.; Hirai, D.; Takeuchi, M.; Ohara, M.; Itai, S.; Iwao, Y. Analysis of available surface area can predict the long-term dissolution profile of tablets using short-term stability studies. *Int. J. Pharm.* **2020**, *586*, 119504. [CrossRef] [PubMed]
34. Higuchi, T. Rate of release of medicaments from ointment bases containing drugs in suspension. *J. Pharm. Sci.* **1961**, *50*, 874–875. [CrossRef] [PubMed]
35. Ritger, P.L.; Peppas, N.A. A simple equation for description of solute release. II. Fickian and anomalous release from swellable devices. *J. Control. Release* **1987**, *5*, 37–42. [CrossRef]
36. Ikeda, S.; Kobayashi, M.; Aoki, S.; Terukina, T.; Kanazawa, T.; Kojima, H.; Kondo, H. 3D-printed fast-dissolving oral dosage forms via fused deposition modeling based on sugar alcohol and poly (vinyl alcohol)—Preparation, drug release studies and in vivo oral absorption. *Pharmaceutics* **2023**, *15*, 395. [CrossRef] [PubMed]
37. Di Cagno, M.P.; Clarelli, F.; Våbenø, J.; Lesley, C.; Rahman, S.D.; Cauzzo, J.; Stein, P.C. Experimental determination of drug diffusion coefficients in unstirred aqueous environments by temporally resolved concentration measurements. *Mol. Pharm.* **2018**, *15*, 1488–1494. [CrossRef] [PubMed]

Journal of
Manufacturing and Materials Processing

MDPI

Article

Geometric and Mechanical Properties of Ti6Al4V Skeletal Gyroid Structures Produced by Laser Powder Bed Fusion for Biomedical Implants

Cong Hou [1,*], Max Goris [1], Dries Rosseel [1], Bey Vrancken [1,2] and Kathleen Denis [1,*]

[1] Department of Mechanical Engineering, KU Leuven, Celestijnenlaan 300, 3001 Leuven, Belgium; max.goris@hotmail.com (M.G.); dries.rosseel@outlook.com (D.R.); bey.vrancken@kuleuven.be (B.V.)
[2] Flanders Make@KU Leuven, 3001 Leuven, Belgium
* Correspondence: cong.hou@kuleuven.be (C.H.); kathleen.denis@kuleuven.be (K.D.)

Abstract: Skeletal gyroid structures possess promising applications in biomedical implants, owing to their smooth and continuously curved surfaces, open porosity, and customisable mechanical properties. This study simulated the geometric properties of Ti6Al4V skeletal gyroid structures, with relative densities ranging from 1.83% to 98.17%. The deformation behaviour of these structures was investigated through a combination of uniaxial compression tests and simulations, within a relative density range of 13.33% to 50% (simulation) and 15.19% to 41.69% (experimental tests). The results established explicit analytical correlations of pore size and strut diameter with the definition parameters of the structures, enabling precise control of these dimensions. Moreover, normalised Young's modulus (ranging from 1.05% to 20.77% in simulations and 1.65% to 15.53% in tests) and normalised yield stress (ranging from 1.75% to 17.39% in simulations and 2.09% to 13.95% in tests) were found to be power correlated with relative density. These correlations facilitate the design of gyroid structures with low stiffness to mitigate the stress-shielding effect. The presence of macroscopic 45° fractures in the gyroid structures confirmed that the primary failure mechanism is induced by shear loads. The observed progressive failure and plateau region proved the bending-dominant behaviour and highlighted their excellent deformability. Additionally, the anisotropy of gyroid structures was confirmed through variations in stress and strain concentrations, deformation behaviour, and Young's modulus under different loading directions.

Keywords: gyroid structure; porosity; Ti6Al4V; laser powder bed fusion; Young's modulus; anisotropy

1. Introduction

Lattice structures are widely recognised for their excellent properties. Among various lattice structures, gyroid structures, a unique type of triply periodic minimal surface (TPMS) lattice structure, are defined by implicit mathematical equations. These structures can be categorised into skeletal gyroid structures, created by filling the void between the TPMS, and sheet-based TPMS lattice structures, where a thickness is assigned to each TPMS [1]. Skeletal gyroid structures exhibit promising potential in biomedical implant applications. Adjusting the parameters in the mathematical equations [2] allows for easy tuning of relative density and mechanical properties. In contrast to conventional strut-based lattice structures, prone to stress concentration at strut junctions, the smooth and continuous TPMS surfaces facilitate seamless geometric transition and provide biomorphic advantages for cells [3]. For the remainder of this paper, skeletal gyroid structures will be referred to as gyroid structures.

The manufacturing of complex geometries like gyroid structures typically involves additive manufacturing technologies. Laser powder bed fusion (LPBF) is one of the most commonly used processes for the additive manufacturing of metal structures. However, defects such as internal pores exist in LPBF-manufactured structures [4,5], which can negatively affect their mechanical properties. Therefore, it is essential to apply suitable process parameters and scan strategies [6], especially as these may differ for porous structures compared to bulky parts.

Healthy human bone needs sufficient stimuli to maintain its density and strength. Without sufficient stimulation, the stress-shielding effect can occur. For instance, when using stiff metal implants, most of the load is borne by the implant, rather than distributed to the surrounding bone. This could lead to bone resorption and weakening over time [7]. Applying gyroid structures in biomedical implants can help mitigate the stress shielding effect and promote osseointegration. The low stiffness of porous gyroid structures shows promise in mitigating this problem. As reported by Ataee et al. [8], the stiffness of Ti6Al4V gyroid structures is comparable to the range of trabecular bone when the relative density ranges from 15 to 18%. Musthafa et al. [9] found that the compressive behaviour of TPMS structures is strongly linked to the design and geometric properties, suggesting that the stiffness can be lowered to approximate that of trabecular bone. Furthermore, the excellent fatigue properties of skeletal gyroid structures make them well-suited for cyclic physiological loading in vivo [10]. Given the complex physiological loading conditions, the design of gyroid structures for implants must consider their mechanical behaviour, particularly stiffness and failure mode.

Gyroid structures feature well-interconnected porosities, which facilitate body fluid flow and potentially enhance bone integration [11]. Hameed et al. [12] investigated gyroid structures with pore sizes ranging from 250 to 400 μm and found these structures to be biocompatible and conducive to stem cell adhesion and spread. To determine the pore and strut sizes, Walker et al. [13] proposed polynomial-fit-based formulas for calculation. Given the importance of relative density and pore size in mechanical properties and biomedical implant applications, it is essential to establish clear and accurate mathematical descriptions for these geometric properties. However, in most studies, the dimensions of geometric features, such as pore and strut size, are typically measured from samples or determined using statistical fits. To the authors' knowledge, explicit equations for precise control of these dimensions still lack investigation.

To deepen the understanding of gyroid structures for biomedical implants, an extended range of geometries is analysed to develop mathematical descriptions of the pores and strut sizes correlated with the definition parameters of gyroid structures. Additionally, both quasi-static compression tests and corresponding simulations are conducted to analyse the mechanical properties and identify the correlations between Young's modulus, yield stress, and relative density. Lastly, the deformation behaviour under different loading directions is investigated. The results of this research provide design information for the rapid determination of parameters in gyroid equations corresponding to specific geometric features. Furthermore, the findings concerning the deformation behaviour of gyroid structures contribute to expanding their potential applications in various fields.

2. Materials and Methods

2.1. Design and Geometric Characterisation of Gyroid Structures

The mathematical equation for gyroid structures was initially defined in 1970 [14] and is given by:

$$F(x, y, z) = cosx \cdot siny + cosy \cdot sinz + cosz \cdot sinx + C$$

where $x = \frac{2\pi X}{U}$, $y = \frac{2\pi Y}{U}$, $z = \frac{2\pi Z}{U}$, (X, Y, Z) represent the coordinates in 3D space. U denotes the unit cell size, and the variable C controls the relative density (volume fraction). A Matlab script was created to generate gyroid structures with various relative densities, unit cell sizes, and boundary sizes. The generated structures were then exported as STL files for further analysis.

Gyroid structures with different C values (−1.5 to 1.5) and unit cell sizes (1 to 4 mm) were created using the Matlab script for geometric property characterisations. These structures were then analysed using Siemens NX (version 2206, Siemens AG, Munich, Germany).

Figure 1a shows the definition of the pore core (green sphere), pore channel, and strut diameter. The geometric data were analysed using an R script [15]. An analytical analysis was performed to obtain the mathematically described geometric properties.

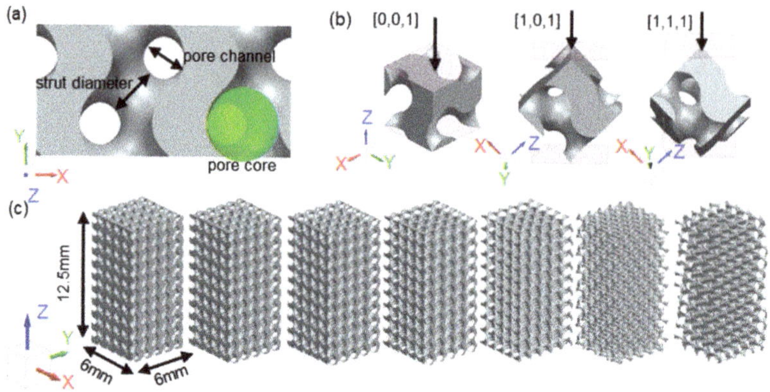

Unit cell size	1.2 mm						
C	0	0.3	0.5	0.8	1.1	1.1	1.1
ρ_{CAD}	50%	40.29%	33.77%	23.80%	13.42%	13.37%	13.33%
Loading direction	[0,0,1]	[0,0,1]	[0,0,1]	[0,0,1]	[0,0,1]	[1,0,1]	[1,1,1]

Figure 1. (**a**) Definition of pore core (green sphere), pore channel, and strut diameter in a two-unit structure. (**b**) Indications of different loading directions based on single units. (**c**) Gyroid structures for compression tests and simulations, with dimensions and loading directions indicated.

For simulation and compression testing, gyroid structures with a unit cell size of 1.2 mm were defined within a boundary volume measuring width = length = 6 mm and height = 12.5 mm (Figure 1c). Four different C values (0.3, 0.5, 0.8, and 1.1) were used for the structures in compression tests and five (0, 0.3, 0.5, 0.8, and 1.1) for simulations. The corresponding theoretical relative density (ρ_{CAD}) of the gyroid structures is presented in Figure 1c.

To investigate the effect of the loading direction on mechanical properties, gyroid structures with C = 1.1 were generated in three distinct orientations, i.e., the default orientation [0,0,1], a rotation along the Y-axis by $-45°$ ([1,0,1]), and a further rotation along the X-axis by 45° ([1,1,1]); see Figure 1b,c.

2.2. Numerical Analysis of Compressive Behaviour of Gyroid Structures

Multistep nonlinear simulations were conducted using a Nastran solver to investigate the mechanical properties of gyroid structures under compressive loading. Gyroid structures of five relative densities (Figure 1c) and three loading directions (Figure 1b,c) were used. A dummy material with a Young's modulus of 110 GPa and a maximum stress of 1300 MPa was used to represent the LPBF-manufactured Ti6Al4V [16]. To determine the element size in simulations, a mesh sensitivity analysis was performed on a gyroid structure (unit cell size 1.2 mm, C = 0.3, ρ_{CAD} = 40.29%, length = width = 2.4 mm, height = 4.8 mm). The subsequent simulations employed quadratic tetrahedral meshes, with the mesh size equivalent to 40% of the strut diameter. Fixed constraints in the Z-axis were applied to the bottom surface of structures. A vertical load corresponding to 2.5% engineering strain was applied to the top surface.

The displacement and reaction force results were then converted into stress–strain curves, further determining the Young's modulus and yield stress of gyroid structures.

Subsequently, the obtained Young's modulus and yield stress values were fit into power relationships correlated with relative density.

The Gibson–Ashby correlation can be used to determine the Young's modulus of porous structures [17,18]. Researchers have reported that the Young's modulus and yield stress of gyroid structures are power correlated with relative density [19,20]:

$$\frac{E}{E_s} = a_1 \left(\frac{\rho}{\rho_s} \right)^{b_1}$$

$$\frac{\sigma_y}{\sigma_{ys}} = a_2 \left(\frac{\rho}{\rho_s} \right)^{b_2}$$

where E, σ_y, and ρ are Young's modulus, yield stress, and density of porous structures, respectively; E_s, σ_{ys}, and ρ_s are Young's modulus, yield stress, and density of solid matrix material, respectively; a_1, a_2, b_1, and b_2 are constants. The normalised Young's modulus and yield stress are defined as E/E_s and σ_y/σ_{ys}, respectively. The relative density (ρ_r) is defined as ρ/ρ_s.

2.3. Manufacturing of Gyroid Structures

A ProX 320B (3D Systems, Leuven, Belgium) LPBF machine was used to manufacture Ti6Al4V-ELI (grade 23, Carpenter Additive, Widnes, UK, particle size 15–53 µm) gyroid samples. Table 1 shows the process parameters employed in the LPBF process. Three specimens were manufactured for each structure. All specimens were then removed from the build plate using wire Electric Discharge Machining (EDM). Subsequently, the specimens underwent cyclic ultrasonic cleaning in water and ethanol to remove residual loose powder.

Table 1. Laser powder bed fusion (LPBF) parameters used in this study.

Laser parameters	Laser power	45 W
	Laser scan speed	300 mm/s
	Laser beam diameter	90 µm
Scan strategy	Layer height	30 µm
	Number of contours	1
	Contour offset	50 µm
	Fill offset	25 µm

The mass and dimensions of the specimens were measured to calculate their structural relative density (ρ_{sample}). The relative density of the struts (built volume) was measured based on Archimedes' principle. The volume of the struts in the gyroid structures was determined by immersing the samples in ethanol and measuring the weight difference in air and ethanol. This difference corresponds to the ethanol weight that occupies the same volume as the struts. Subsequently, the weight of the samples was divided by the measured volume of the struts to calculate their strut density. The relative density of struts is calculated using the density of solid Ti6Al4V (4.43 g/cm^3) as the reference. Three solid specimens, following the same procedures of manufacturing and postprocessing, were produced as the control group for compression tests.

2.4. Compression Test

Quasi-static compression tests were conducted using an Instron 3367 testing system (Instron, Norwood, MA, USA), with a preload of 50 N at a 0.5 mm/min rate. Subsequently, specimens were compressed at a constant engineering strain rate of 0.001 /s. During tests, the strain was continuously measured using the Video Extensometer (Instron AVE2, Instron, Norwood, MA, USA) and the deformation of the specimens was recorded using a 12M-pixel camera (IMX 603, Sony, Tokyo, Japan). The Young's modulus and yield stress for each specimen were calculated based on the stress–strain curves in the compression test data, which were further fit into power relationships and compared with the correlations derived from simulations.

To enable the direct comparison of samples that experienced large deformation, the specific absorbed energy is limited to the strain of 30%:

$$W = \int_0^{0.3} \sigma(\varepsilon) d\varepsilon$$

where W is specific absorbed energy, σ and ε are engineering stress and strain during compression tests, respectively.

3. Results

3.1. Geometric Properties of Gyroid Structures

The relationship between the relative density of gyroid structures and the parameter C exhibits a linear correlation, Figure 2a. The relative density increases with lower C values, leading to thicker struts and smaller pores. A relative density of 50% corresponds to C being 0. C higher than 1.41 leads to non-continuous volumes, while C equals −1.5 generates fully dense structures. Neither struts nor pore channels are visible when C equals or undercuts −1. Examining the strut diameter, pore channel, and pore core sizes displayed as green spheres in Figure 2b–d, the values reveal linear proportionalities to the unit cell size when C falls within the −1 to 1 range. Consistent results were found for strut diameter, pore core, and pore channel sizes when comparing CAD measurements (green spheres) with explicit analytical results (red surfaces).

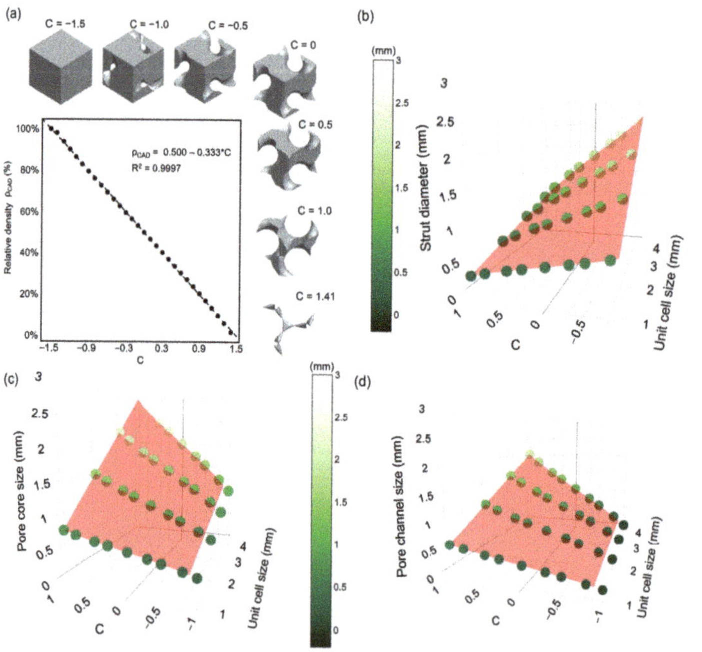

Figure 2. Geometric characterisations of gyroid structures. (**a**) Correlation between relative density and parameter C, with images of the gyroid unit for different C values. Correlations of strut diameter (**b**), pore channel (**c**), and pore core (**d**) with unit cell size U and parameter C. Green spheres represent the feature size measured from CAD models. The red surfaces represent analytical analysis results and can be described as equations by the end of Section 3.1. The detailed results of the geometric analysis can be found in the Supplementary Data.

Figure 3 shows the approach of analytical analysis on feature sizes in gyroid structures. Given the continuous repeating units in gyroid structures, the pore core, pore channel, and strut can be calculated from specific line sections (L_1 to L_4) using the following equations:

$$Pore\ core = L_1 + L_4$$

$$Pore\ channel = 2L_4$$

$$Strut\ diameter = \sqrt{2}\frac{U}{2} - 2L_4$$

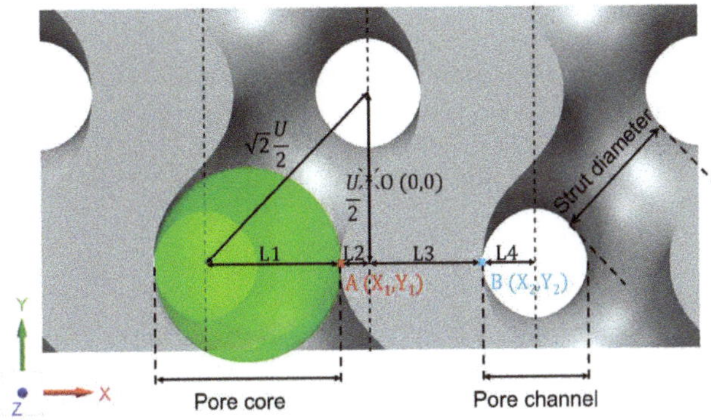

Figure 3. Schematic representation of calculating the geometric feature sizes using the analytical approach in a two-unit gyroid structure. The dimensions of this structure are 2 units in X, 1 unit in Y and Z. The centre of the image is defined as the 0 point.

The length of these line sections can be determined by calculating the X coordinates of two points on the gyroid surface (A and B). The gyroid surface equation can be rewritten as

$$F(x,y,z) = cos\frac{2\pi X}{U} \cdot sin\frac{2\pi Y}{U} + cos\frac{2\pi Y}{U} \cdot sin\frac{2\pi Z}{U} + cos\frac{2\pi Z}{U} \cdot sin\frac{2\pi X}{U} + C$$

with $(X,\ Y,\ Z)$ representing the point coordinates in 3D space. X_1 and X_2 can be calculated by setting $Y = -\frac{U}{4}$ (the Y value of line sections L_1 to L_4) and $Z = \frac{U}{2}$ (the front surface in Figure 3). The gyroid surface equation is then simplified as:

$$-\cos\left(\frac{2\pi X}{U}\right) - \sin\left(\frac{2\pi X}{U}\right) + C = 0$$

When C ranges between −1 and 1, the solution for X is:

$$X = \frac{U\left(\tan^{-1}\left(\frac{1\pm\sqrt{2-C^2}}{C+1}\right) + \pi n\right)}{\pi}, when\ C+1 \neq 1\ and\ n \in Z$$

Considering πn is repeating in 3D space, the solution can be further simplified as:

$$X_1 = \frac{U\left(\tan^{-1}\left(\frac{1-\sqrt{2-C^2}}{C+1}\right)\right)}{\pi}, X_2 = \frac{U\left(\tan^{-1}\left(\frac{1+\sqrt{2-C^2}}{C+1}\right)\right)}{\pi}$$

The line sections can be calculated from X_1 and X_2:

$$L_1 = \frac{U}{2} + X_1, L_2 = -X_1,\ L_3 = X_2,\ L_4 = \frac{U}{2} - X_2$$

The geometric features can be determined as follows:

$$Pore\ core = L_1 + L_4 = U + X_1 - X_2 = U + \frac{U\left(\tan^{-1}\left(\frac{1-\sqrt{2-C^2}}{C+1}\right)\right)}{\pi} - \frac{U\left(\tan^{-1}\left(\frac{1+\sqrt{2-C^2}}{C+1}\right)\right)}{\pi};$$

$$Pore\ channel = 2L_4 = U - \frac{2U\left(\tan^{-1}\left(\frac{1+\sqrt{2-C^2}}{C+1}\right)\right)}{\pi};$$

$$Strut\ diameter = \sqrt{2}\frac{U}{2} - 2L_4 = \sqrt{2}\frac{U}{2} - U + \frac{2U\left(\tan^{-1}\left(\frac{1+\sqrt{2-C^2}}{C+1}\right)\right)}{\pi};$$

The analytical results of strut diameter, pore channel, and pore core sizes are visualised as red surfaces in Figure 2b–d.

3.2. Uniaxial Compressive Behaviour of Gyroid Structures Through Simulation

Figure 4 shows the results of the mesh sensitivity analysis. The convergence of Young's modulus, indicated by a less than 1% difference between results, is achieved when the element size of the structures is 40% of the strut diameter or less.

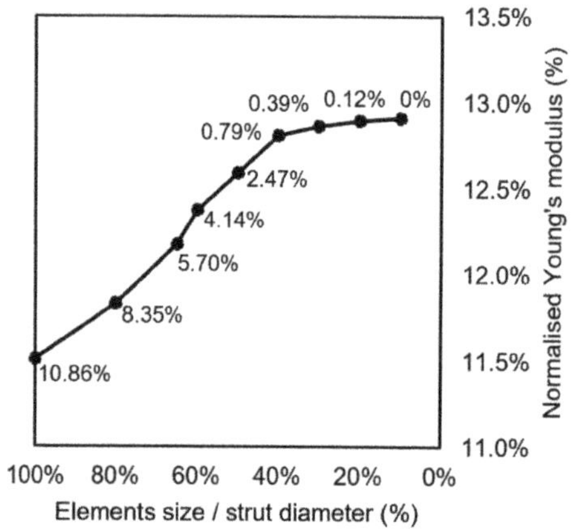

Figure 4. Results of mesh sensitivity analysis based on Young's modulus. Differences between the results for various element sizes and the final values (element size = 10% of strut diameter) are indicated next to the marks.

Figure 5a,b illustrate the stress and strain distributions within gyroid structures at 1% overall strain. Notably, stress and strain concentrations in structures loaded in the [0,0,1] direction appear at a 45° angle relative to the loading direction. The stress concentrations increase with relative density, as indicated by higher stress values in the concentrated regions. Stress and strain also concentrate on the thin features of the structure boundaries (surfaces). The influence of loading direction on stress and strain distribution is evident in the three structures with C = 1.1. In structures loaded in the [1,0,1] direction, stress and strain are concentrated in the middle of horizontal and vertical struts, whereas the concentration in structures loaded in the [1,1,1] direction appears in the middle of struts oriented at nearly 45°. Furthermore, strain concentration areas indicate the potential failure points or zones prone to deformation, highlighting critical regions that could lead to structural instability under further loading.

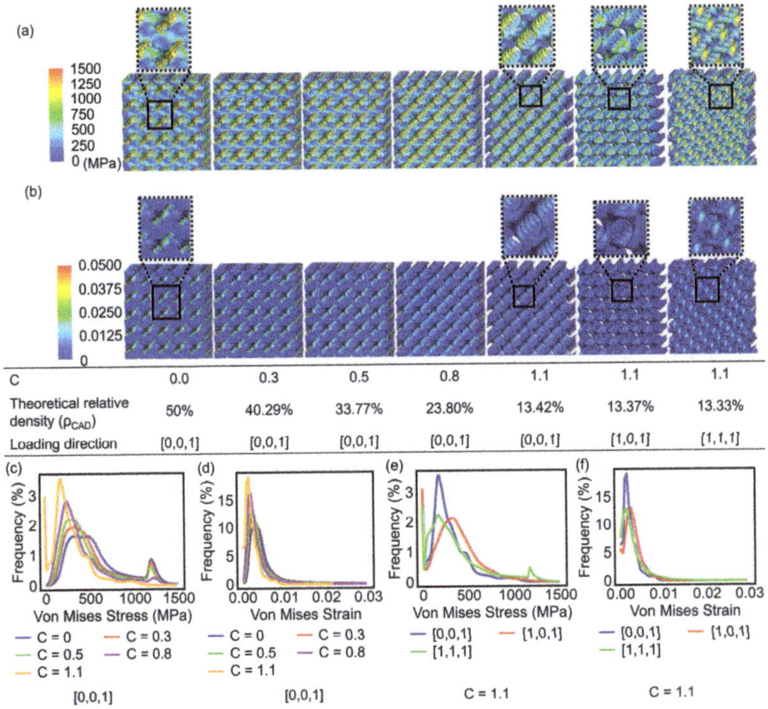

Figure 5. Simulation results of stress distributions (**a**) and strain distribution (**b**) in gyroid structures at 1% overall strain (before yield), with indications of parameter C, relative density, and loading direction. Higher magnification of stress and strain distributions are shown. Corresponding histograms are shown for stress (**c**) and strain (**d**) in gyroid structures with varying relative density (loading direction [0,0,1]), and stress (**e**) and strain (**f**) in gyroid structures with different loading directions (C = 1.1).

The histograms in Figure 5c to f further illustrate these patterns. In structures loaded in the [0,0,1] direction, stress and strain shift towards higher values with increased relative density (lower C). When comparing the three loading directions in the range of high stress (over 1000 MPa) and strain (over 0.01), the structure loaded in the [1,1,1] direction exhibited the highest values, while the structures loaded in the [0,0,1] direction showed the lowest values. Furthermore, similar to the denser structures, the structure loaded in the [1,1,1] direction displayed a stress peak at around 1200 MPa in its stress distribution histogram (Figure 5d).

The coloured curves in Figure 6 present the stress–strain curves derived from simulation results. All structures demonstrated linear elastic sections, yield, and stress-hardening in the curves. In the comparison of relative densities (Figure 6a), porous structures exhibit lower stress values and flatter slopes in the linear elastic sections. When comparing the linear sections in curves of different loading directions, the structure loaded in the [1,1,1] direction showed the highest stress and slope, while the [0,0,1] direction exhibited the lowest (Figure 6b). Table 2 includes the simulation results of Young's modulus and yield, both of which show a positive correlation with relative density.

Table 2. Properties of gyroid structure in simulations and compression tests.

C	Loading Direction	Theoretical Relative Density (ρ_{CAD})	Young's Modulus (Simulation)	Yield Stress (Simulation)	Relative Density of the Samples (ρ_{sample})	Relative Density of the Struts	Young's Modulus (Test)	Yield Stress (Test)	Maximum Stress (Test)	Strain at First Maximum Stress (Test)	Specific Absorbed Energy (Test)
Unit	-	[%]	[GPa]	[MPa]	[%]	[%]	[GPa]	[MPa]	[MPa]	[%]	[MJ/m^3]
control	-	100	110	1150	99.53 ± 0.12	-	78.88 ± 3.19	979.59 ± 2.07	1370.67 ± 13.38	16.41 ± 0.25	195.01 ± 4.52
0	[0,0,1]	50.00	22.85	199.99	-	-	-	-	-	-	-
0.3	[0,0,1]	40.29	14.07	131.33	41.69 ± 0.14	98.96 ± 0.06	12.25 ± 0.06	136.68 ± 0.67	191.54 ± 1.37	6.07 ± 0.15	46.62 ± 4.87
0.5	[0,0,1]	33.77	9.47	94.80	35.30 ± 0.21	99.17 ± 0.22	7.99 ± 0.56	94.75 ± 4.09	136.70 ± 0.56	6.46 ± 0.06	34.44 ± 0.40
0.8	[0,0,1]	23.80	4.27	50.86	25.72 ± 0.18	99.20 ± 0.14	3.72 ± 0.04	52.93 ± 1.09	75.81 ± 0.32	6.30 ± 0.17	19.00 ± 0.23
1.1	[0,0,1]	13.42	1.15	20.12	15.46 ± 0.17	99.31 ± 0.16	1.30 ± 0.01	20.44 ± 0.42	29.58 ± 0.31	6.26 ± 0.14	6.87 ± 0.10
1.1	[1,0,1]	13.37	1.85	31.19	15.19 ± 0.19	99.33 ± 0.26	2.18 ± 0.05	28.78 ± 0.41	37.27 ± 0.57	3.64 ± 0.48	7.49 ± 0.10
1.1	[1,1,1]	13.33	2.31	26.37	15.97 ± 0.02	99.28 ± 0.25	2.03 ± 0.08	25.11 ± 0.11	34.39 ± 0.82	4.36 ± 0.42	7.71 ± 0.11

Young's modulus and yield stress derived from simulation results are power correlated with relative density, Figure 7. These power relationships demonstrate high R^2 values, yet they differ from the results reported by Gibson–Ashby [18] and Yan et al. [20].

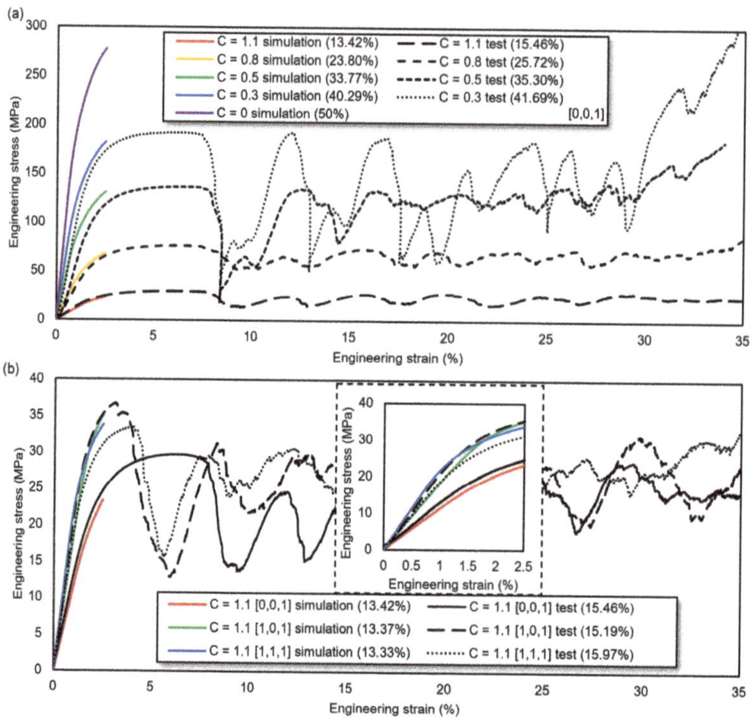

Figure 6. Comparison of stress–strain curves of gyroid structures with (**a**) different relative density (loading direction [0,0,1]) and (**b**) loading directions (C = 1.1). Loading directions are indicated in square brackets. The corresponding ρ_{CAD} and the average ρ_{sample} are shown in round brackets.

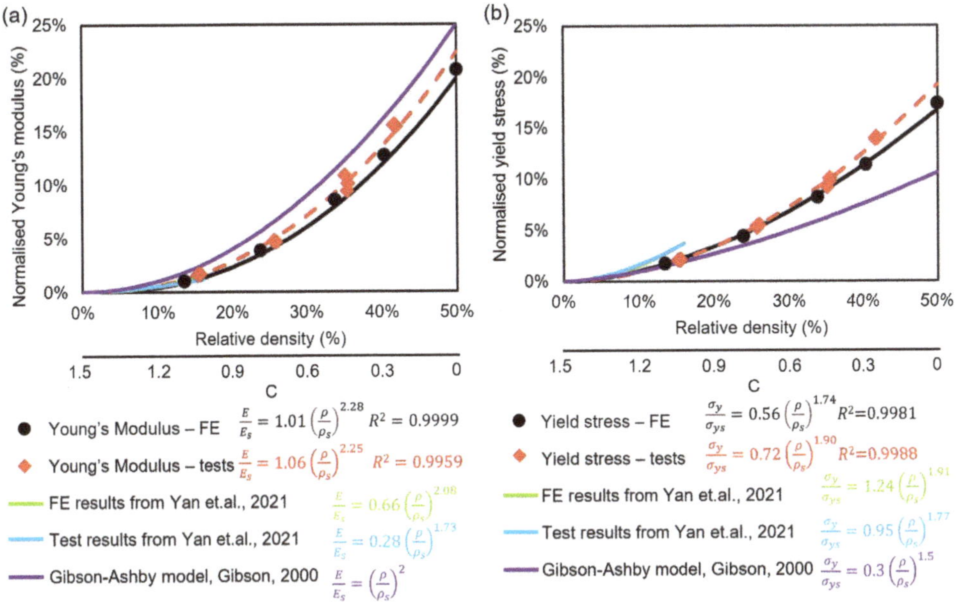

Figure 7. (**a**) Power relationships between normalised Young's modulus and relative density of gyroid structures. (**b**) Power relationships between normalised yield stress and relative density of gyroid structures. Structures are loaded in the [0,0,1] direction. The theoretical relative density (ρ_{CAD}) was used to represent data in simulations, and the measured relative density (ρ_{sample}) was used to illustrate data in tests. The power relationships are compared with the work of Yan et al. [20] and the Gibson–Ashby [18] model.

3.3. *Verification of Mechanical Behaviour for Gyroid Structures Through Compression Tests*

Figure 8 shows the optical microscope images of samples after LPBF manufacturing; a large number of adherent particles are visible, indicating a slight overbuild. Table 2 presents the measured relative density (ρ_{sample}) of samples. Notably, all samples demonstrated a higher relative density than their theoretical values, with most samples exhibiting a relative density of struts exceeding 99%.

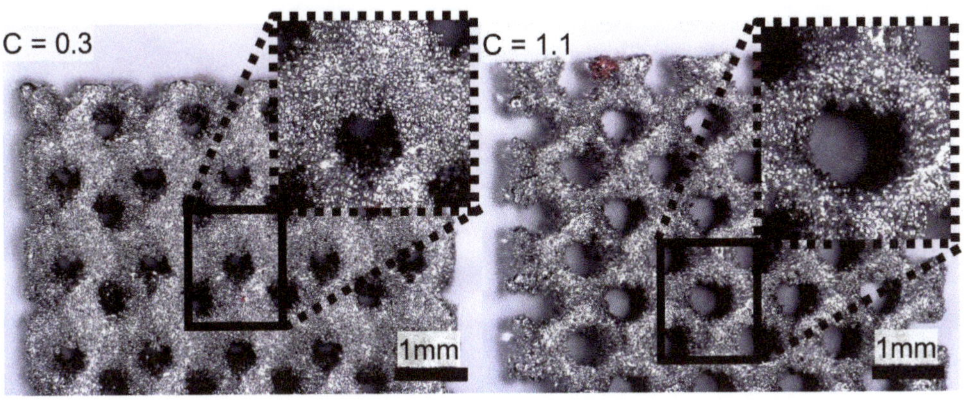

Figure 8. Optical microscope images of as produced samples.

Figure 6a shows the stress–strain curves obtained from the quasi-static compression tests of samples with different relative densities. The gyroid structures in the compression tests exhibited curves similar to simulation results up to a strain of 2.5%, exhibiting linear elastic sections, yield, and strain-hardening stages. Dense structures demonstrated higher Young's modulus and yield stress than porous structures. Initial fractures of all samples loaded in the [0,0,1] direction occurred at a strain of approximately 8%. Table 2 presents the relative density and mechanical properties of gyroid samples in the compression tests. Among the structures loaded in the [0,0,1] direction, Young's modulus, yield stress, maximum stress, and specific absorbed energy exhibit positive correlations with relative density. In these structures, the strain at the first maximum stress is approximately between 6% and 6.5%. However, the two structures loaded in different directions display higher Young's modulus, yield stress, maximum stress, lower strain at first maximum stress, and comparable specific energy (Figure 6b and Table 2). Compared to the structures loaded in the [0,0,1] direction, the structures loaded in the [1,0,1] and [1,1,1] directions showed earlier initial fractures at higher stress, along with greater differences in the stress–strain curves of simulations and tests. Nevertheless, plateau regions, seen as fluctuating stress over large strain ranges, were identified after initial failure in all samples. Increased stress due to structural densification was observed towards the end of the test. The impact of relative density and loading direction on mechanical properties will be further discussed in Sections 4.2 and 4.3.

Table 3 further confirms the variation in Young's modulus when the gyroid structures were subjected to different loading directions. The simulation results indicate the highest Young's modulus in the [1,1,1] direction and the lowest values in the [0,0,1] direction, aligning with the findings of Yan et al. [20]. However, the Young's modulus derived from tests exhibits higher values and demonstrates a different pattern across loading directions. Unlike the simulation results where the structures loaded in the [1,1,1] direction have the highest Young's modulus, the structures loaded in the [1,0,1] direction display the greatest stiffness in tests.

Table 3. Comparison of normalised Young's modulus in this study and calculated values based on the same relative density (ρ_{CAD}) from Yan et al. [20].

C	Loading Direction	Result of This Study		Yan et al. [20]
		Simulation	Test	FE
1.1	[0,0,1]	1.05%	1.65 ± 0.02%	1.12%
	[1,0,1]	1.68%	2.76 ± 0.06%	1.56%
	[1,1,1]	2.10%	2.57 ± 0.10%	2.10%

Table 4 details the progressive failures observed in gyroid structures during compression tests. Most specimens exhibited concentrated initial failures at approximately 45° relative to the loading direction, followed by progressive failures and eventual densification. Dense structures (C = 0.3) underwent horizontal shifts while porous structures (C = 1.1) experienced densification with vertical stacking post-failure. During deformation, struts bent, fractured, and gradually contacted each other, indicating a bending-dominant deformation and failure mode. The initial fracture locations in struts corresponded to stress and strain concentrations. Specifically, failures in the structures loaded in the [0,0,1] direction initiated from the 45°-oriented struts, while in the structures loaded in the [1,0,1] direction, failures initialised in the middle of struts. Conversely, in the structures loaded in the [1,1,1] direction, failure involved the bending of numerous 45°-oriented struts.

Table 4. Deformation and failure mode of gyroid samples during compression tests.

C	ρ_{sample} (%)	Loading Direction	Strain (ε, %)				
0.3	41.69 ± 0.14	[0,0,1]	0	1.5	11.1	20.4	34.0
			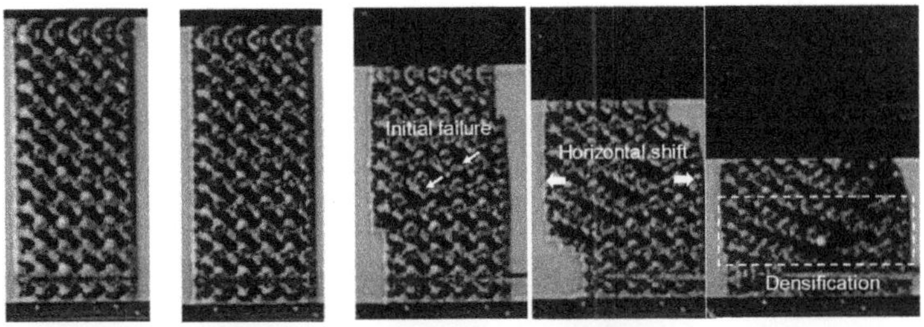				
1.1	15.46 ± 0.17	[0,0,1]	0	1.8	11.8	20.4	38.0
			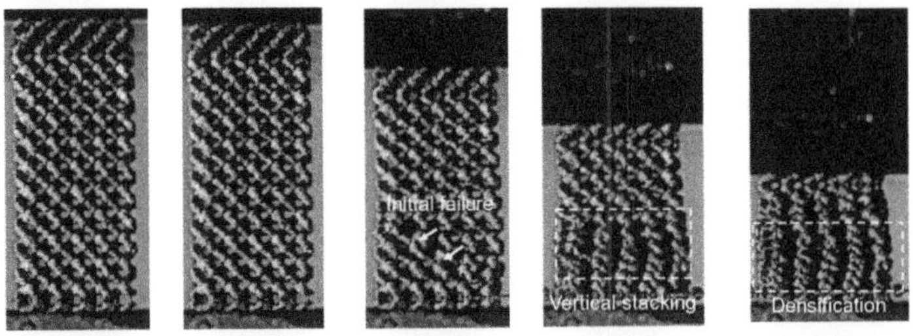				
1.1	15.19 ± 0.19	[1,0,1]	0	2.3	14.8	30.0	37.0
			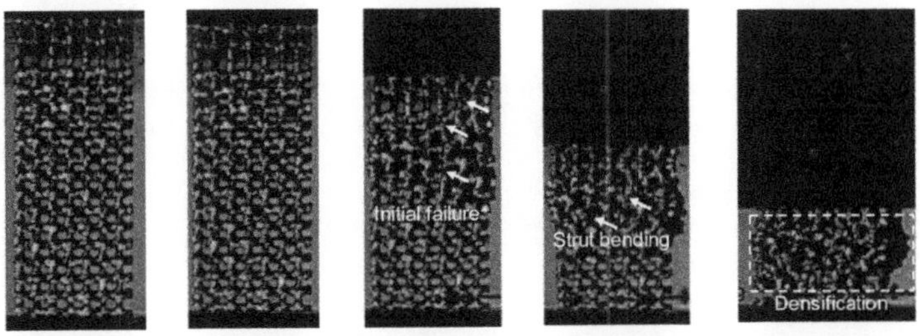				
1.1	15.97 ± 0.02	[1,1,1]	0	2.4	11.1	24.6	40.0
			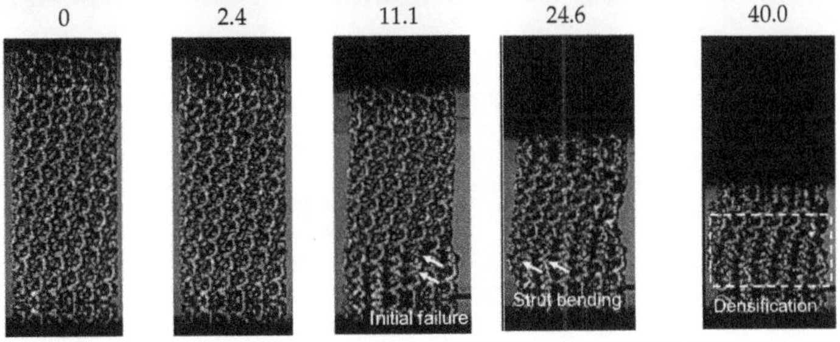				

4. Discussion

4.1. Adaptive Geometric Properties

The geometric properties of gyroid structures are strongly associated with the parameter C and the unit cell size. The observed linear relationship between relative density and C aligns with the findings of other researchers [2,21]. Higher C values decrease the relative density of gyroid structures, leading to thinner struts and larger pores. Maintaining a continuous volume within the porous gyroid structure typically requires a C value ranging from −1.5 to 1.41. Pore channels and struts are not visible in gyroid structures with C values lower than −1.0, thus falling within the applicable C value range of −1.0 to 1.41. In any case, pore and strut sizes follow the analytical relationship within the C value range of −1.0 to 1.0. With a fixed C value, gyroid structures with varying unit cell sizes maintain identical geometries, with their dimensions scaling proportionally. Consequently, strut diameter, pore channel, and pore core sizes exhibit linear correlations with the unit cell size, as evidenced by both CAD measurements and explicit analyses. These explicit equations facilitate precise control over geometric features, enabling rapid determination of design parameters (unit cell size and C) to generate gyroid structures with specific relative density, pore size, and strut diameter.

The high porosity of gyroid structures holds promise for biomedical orthopaedic implant applications, offering enhanced osseointegration and improved implant stability. The continuous and non-angular TPMS sustains an ideal environment for cell attachment and proliferation [3]. Pores are crucial in facilitating these cellular activities [22]. Interconnected pores allow fluid flow, promoting nutrient and oxygen supply for cells [23,24]. Porous materials with a relative density below 50% [23,25,26] and pores sizes ranging from 100 to 700 μm [22,27] have demonstrated considerable potential for osseointegration. The correlations derived from this geometric analysis guide the determination of suitable ranges for C and unit cell size to generate gyroid structures with high potential in osseointegration, particularly by selecting C values greater than 0 and using specific combinations of C and unit cell size.

4.2. The Effect of Relative Density on Compressive Behaviours

In general, the employed LPBF process parameters ensured satisfactory manufacturing quality of the samples. The high relative densities of the struts confirm the low ratios of porosity defects. Yet, the slightly higher relative density (ρ_{sample}) observed in samples is usually characterised as overprinting, due to heat input during the process. The additional heat inevitably affects the material adjacent to the built volume. Consequently, particles may adhere to the sample surface, increasing both built volume and relative density.

The stress and strain distributions in simulations reveal the deformation behaviour of gyroid structures under compressive load. Struts in gyroid structures experience a combination of bending and compression deformation, with bending being the main cause of stress concentration and subsequent failures [10,28]. Stress and strain concentration predominantly occur on struts oriented at a 45° angle relative to the loading direction, corresponding to the direction of the maximum shear stress. The histograms reflect the intense stress concentrations in dense structures. Most materials experience low stresses, seen as the high-frequency peaks at lower stress ranges. In contrast, the material on 45° orientated features experiences higher stress and displays low-frequency peaks at higher stress ranges. At equal engineering strain, the stress concentration is more pronounced in dense structures, attributed to their higher stiffness. The observations suggest that increased relative density leads to greater stress concentration within the structures, which aligns with the findings reported by Yang et al. [10]. Stress concentration approaches the yield stress of Ti6Al4V, implying the initial failure locations at stress concentrations. The thin features on the boundary surfaces of the structures caused unfavourable stress and strain concentrations. These locations are highly susceptible to initial failures, potentially leading to numerous sharp edges and ends that could pose risks in biomedical applications. Hence, implant designs should aim to minimise or avoid such thin features.

The stress–strain curves illustrate distinct deformation behaviours of gyroid structures with varying relative densities. In general, denser structures showed higher stress values due to their high stiffness. Both simulation and test curves consistently demonstrate the linear elastic sections and yield, confirming that the material model used in simulations can represent the linear elastic deformation and yield of LPBF-manufactured Ti6Al4V. However, the maximum deformation in simulations is restricted to strain = 2.5% due to software limits, resulting in the exclusion of plateau regions in the simulation curves. Discrepancies between simulation and test curves can be attributed to using a simplified material model in simulations. Furthermore, the wide plateau regions in tests consist of stress peaks and valleys over a large strain range, resulting from local collapse and regained support during continuous deformation. The periodic changes of stress over deformation align with the observation by Fan et al. [29]. This implies the good deformability of gyroid structures during progressive failures, aligning well with their application as energy absorbers.

The deformation behaviours and failure modes of gyroid structures are strongly influenced by relative density. In the structures loaded in the [0,0,1] direction, stress typically concentrates at thin features or sharp corners. Due to the period geometry of gyroid structures, its cross-section reaches the minimum area at 45° with respect to the loading direction [29], corresponding to a weak orientation prone to failures primarily induced by the shear load [30]. The failure manifests as bending, plastic deformation, and fractures of the struts. The relatively thicker features in dense structures enhance local stiffness, enabling the failure and defect accumulations near initial cracks. The high stiffness also ensures that the resilient volume can support the fragments during progressive failure, seen as fragments of dense structures that horizontally shifted after initial fractures. Thus, the dense structures showed 45° diagonal fractures. Conversely, the low stiffness associated with thin struts in porous structures results in localised failure initiation. Throughout the compression tests, the low stiffness of the material cannot horizontally shift the fragments. Thus, the deformed volume in porous structures tends to stack vertically, causing layer-by-layer failures and gradual densifications.

The power relationships of Young's modulus and yield stress show notable consistency between simulation and test results, affirming the robustness of both approaches. The slight difference can be attributed to inherent manufacturing defects within the LPBF process and the slight increase in dimensions and relative density of specimens. Transitioning to a comparison with existing literature, divergence from the power relationships proposed by Gibson–Ashby [18] is understandable, given their use of metallic foam, a porous material distinct from the skeletal gyroid structure. Larger disparities with the work of Yan et al. [20] can be attributed to their focus on more porous gyroid structures with a narrow relative density range (5% to 15%). This limitation can narrow the applicability of formulas that are intended to describe mechanical properties across the entire relative density range.

The power relationships of Young's modulus and yield stress offer high customisability in various applications. By selecting appropriate C values corresponding to the desired relative density, gyroid structures with precise Young's modulus can be accurately generated. This variability highlights the excellent potential of gyroid structures in biomedical implants, where stiffness can be tailored to align with the value of human bone (typically around 20 GPa or lower [31]). As demonstrated in [32], low-stiffness porous materials can be used to replace the volume of stiff metal stem components in artificial joints, thereby reducing the overall stiffness of the stem and allowing the bone to be stimulated to maintain its integrity. Based on the results of this study, gyroid structures with $C = 0$ achieves a stiffness of approximately 20 GPa. This geometric configuration presents a promising solution for reducing the stress-shielding effect. Similar adaptive mechanical properties are observed in various porous materials, which further explores the range of porous materials for biomedical applications [33].

4.3. The Effect of Loading Direction on Compressive Behaviour and Anisotropy

The stress distribution of gyroid structures reflects the loading conditions of the struts. In lattice structures, the struts aligned with the loading direction primarily experience compression during deformation while lateral struts undergo stretching [34]. By using the approach proposed by Yang et al. [19], gyroid structures can be simplified as scaffolds of cylindrical struts, allowing for the computational determination of strut orientations using line beam data. The unique shapes of gyroid lattice structures result in struts orienting across a wide spectrum of angles, see Figure 9. Therefore, struts experience a combination of bending and compression due to this diverse range of orientations. Comparisons of stress distribution (Figure 5e) reveal significant stress concentration in structures loaded in the [1,0,1] and [1,1,1] directions. In the stress range over 1000 MPa and the strain range higher than 0.01, the structures loaded in the [1,1,1] direction consistently display the highest concentration, followed by the structures loaded in the [1,0,1] direction and [0,0,1] direction. Young's modulus in the simulation results of these three structures follows the same sequence (Table 3). This correspondence between stress concentration and stiffness is consistent with the findings from numerical analysis of the structures loaded in the [0,0,1] direction, indicating that stiffer structures demonstrate more pronounced stress concentrations under uniform compressive strain.

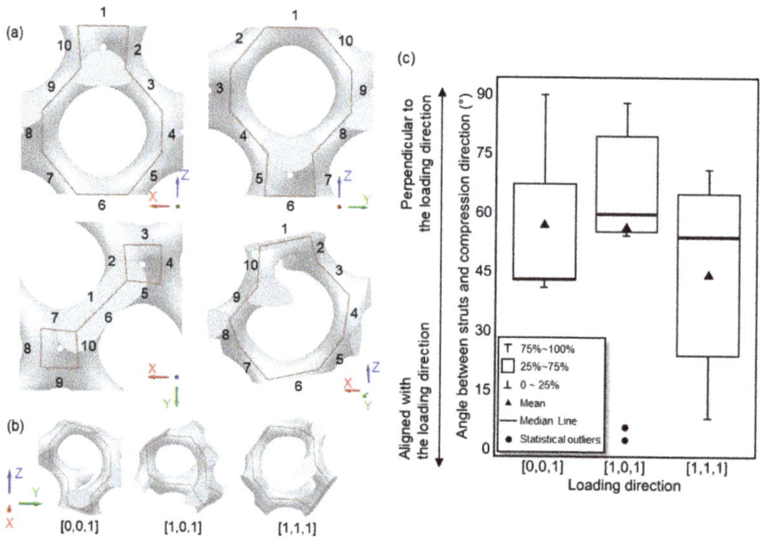

Figure 9. Struts orientation analysis of gyroid unit cell with parameter C = 1.1. (**a**) Schematics of gyroid units in different perspectives. The grey colour represents solid volume and the orange line beams represent the simplified struts. Corresponding labels for struts are shown. (**b**) Schematics of gyroid units in different loading directions. (**c**) The strut orientation for different loading directions, shown as the angle between struts and compression direction (Z-axis).

The mechanical response of gyroid structures changes with loading directions due to the varied strut orientations and distinct load transfer modes. Under compression, the load traverses through the lattice geometry towards the bottom of the structures, with the efficiency of this transfer influenced by the structure's geometry and strut orientation. Cuadrado et al. [35] compared the mechanical properties of lattice structures with different unit cells. Their results showed that as the inclination angle between struts and the loading direction increases, the deformation mechanism transfers from compression-dominant to bending-dominant, consequently reducing mechanical properties. The comparison of strut orientation in gyroid structures (Figure 9) reveals that struts in the structures loaded in the [0,0,1] direction are aligned at larger angles with the loading direction, indicating more bending-dominant behaviour. When the structure is loaded in the [1,0,1] direction, some struts align closely with the loading direction, thereby enhancing the stiffness. However, most struts remain inclined, which restricts the stiffness. Compared to the other two directions, the [1,1,1] direction exhibits the greatest number of struts closely aligned with the compression direction. As a result, the deformation mechanism approximates a compression-dominant behaviour, leading to the highest observed stiffness in the simulation results. This direction-specific stiffness underscores the inherent anisotropy of gyroid structures.

The stress–strain curves also elucidate the influence of loading directions on the mechanical response of gyroid structures. Structures loaded in the [1,0,1] and [0,0,1] directions exhibit increased stiffness and a decreased strain at initial fractures during testing. In these structures, local stress reaches the yield point at a lower strain. In the subsequent plastic deformation, defects such as dislocations, micro-cracks, and grain boundary separations rapidly accumulate at significant strain concentration locations, leading to fractures at lower strains. The increased stiffness observed in these two loading directions is accompanied by corresponding low strain at initial fractures. However, their superior deformability remains unaffected, as evidenced by wide plateau regions and comparable absorbed energy.

The mechanical properties of lattice structures are also highly dependent on the manufacturing process. The broad range of strut orientations presents challenges to the LPBF process. Overhangs can occur when struts align more closely with the build plate. Without a sufficient solid substrate, the new layer of materials in the overhang regions is prone to increased spattering. This results in less solid deposition, increasing defects such as material loss and high surface roughness [36]. Kranz et al. [37] discovered that struts more parallel to the build plate tend to induce defects such as high surface roughness and dimensional inaccuracies. Shange et al. [38] reported increased porosity when features align more closely with the build plates. These defects ultimately compromise the mechanical properties, reducing stiffness and deformability. Therefore, the support ratio of the LPBF process, defined as the proportion of the material layer supported by the previous layer, could indicate manufacturing quality. Figure 10 compares the material layers for manufacturing gyroid structures in three loading directions, revealing significant differences in sequential layer shapes for the three loading directions, associated with different support ratios. To further quantify the support ratios, the CAD models of the structures, oriented in three loading directions, were sectioned using the layer height of the LPBF process. These section images were then compared to calculate support ratios between layers; see Figure 11. Notably, the 2D geometries of the slices used in this analysis are identical to those used in generating laser trajectories.

The Young's modulus in compression tests exhibited higher values than those from the simulation results (Table 3). This increase can be attributed to the slightly higher relative densities (overbuild) of the produced gyroid structures (see Figure 8). Despite the consistent power relationships of Young's modulus in tests and simulations, discrepancies emerge for different loading directions. Structures loaded in the [1,0,1] direction demonstrate significantly higher stiffness in tests than the other two loading directions, which can be attributed to the improved manufacturing quality. Figure 11c illustrates the support ratios of the material layers when structures are produced and loaded in three directions. Among these, the [1,0,1] direction exhibits the highest support ratio. Consequently, this direction is expected to have the fewest overhangs and thus the least manufacturing defects. This results in an improved load-bearing capacity for this loading direction, leading to a higher Young's modulus. In contrast, the relatively low support ratios in the other two directions increase manufacturing defects and reduce Young's modulus, as evidenced in Table 3. Defects, such as porosity, can be significantly reduced by optimising laser parameters and scan strategies in the LPBF process, including adjustments to laser power and scan strategy, as demonstrated in a previous work of the authors [39].

The failure modes of gyroid structures under the three loading directions correspond to the stress concentrations observed in simulations. The initial failure locations of structures align with the strain concentration, specifically in the middle of struts where stress and strain concentrations are highest. The struts that align more closely with the loading direction, particularly in the [1,0,1] and [1,1,1] directions, exhibit a compression-dominant deformation. However, given the wide range of strut orientations, most struts display a bending-dominant deformation mechanism. The primary failure mode in gyroid structures involves the bending and rotation of struts, indicating a bending-dominant behaviour. During the progressive failure, deformed and failed struts stacked together due to the low stiffness of the porous structures (with a relative density of around 15%) were unable to support the resilient volume.

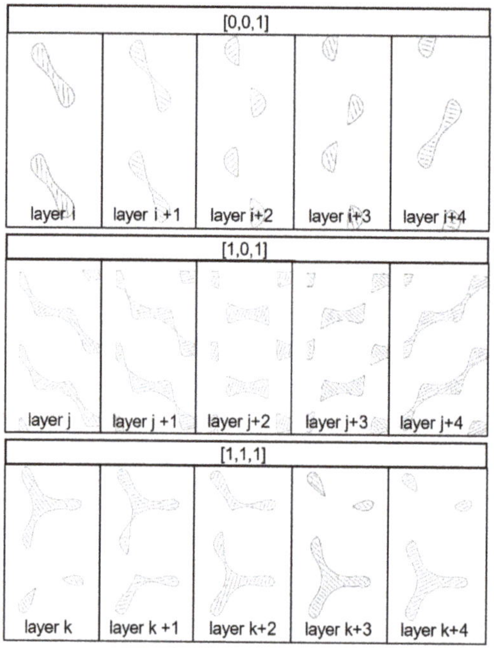

Figure 10. Detailed view of sequential building layers in slicing gyroid structures for three loading directions. Parameter C = 1.1.

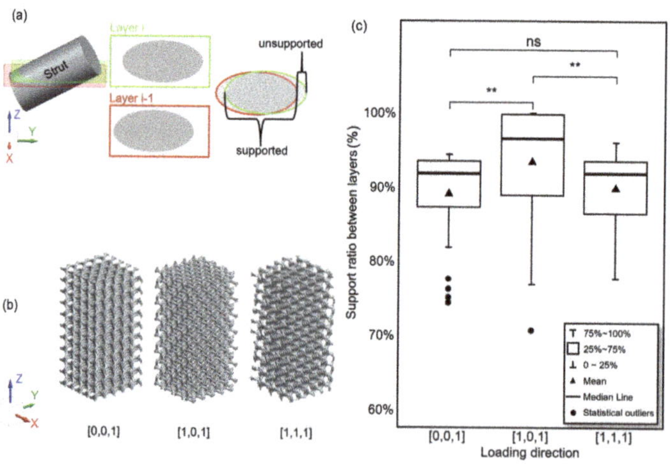

Figure 11. (**a**) Schematic of the principle of calculating the support ratio of material layers. (**b**) CAD models of gyroid structures in FEM analysis and compression tests, corresponding to Figure 1c. (**c**) Box plots of support ratios of gyroid structures in three loading directions. Comparisons show different significance codes. p-values ranges: 0.001 < ** < 0.01, 0.05 < ns (not significant).

The loading direction-dependent mechanical properties confirm the anisotropy of gyroid structures and expand the design possibilities for tailoring these structures to specific applications. In instances where a structure meets geometric parameter requirements, such as pore sizes, but requires increased stiffness, optimising the orientation of the gyroid structure can be an effective design strategy. Moreover, the flexibility and potential of gyroid structures can be significantly enhanced when integrated with a graded structure design. The radial gradient design, as demonstrated in [21], is a promising approach to expanding the design possibilities of graded structures for enhancing mechanical performance and biological integration. For orthopaedic implants, the surface of an implant can be substituted with finely tuned gyroid structures exhibiting non-uniform stiffness and pore size. Consequently, the stress distribution at the bone interface can be customised to more effectively mitigate the stress-shielding effect, and the osseointegration process can be improved by applying an appropriate pore size.

5. Conclusions

This study investigated the geometric characteristics and the quasi-static compressive behaviour of skeletal gyroid structures. The key findings highlight the explicit description and control of gyroid structure geometries by adjusting parameters in mathematical definitions. The results demonstrated that increasing relative density elevates both Young's modulus and yield stress. Moreover, this study revealed the anisotropic behaviour of gyroid structures under various loading directions.

The relative density of gyroid structures exhibits a linear correlation with the parameter C. Pore size and strut diameter follow explicit equations based on C and unit cell size. The equations empower the precise generation of gyroid structures with specified relative density, pore size, and strut diameter.

The mechanical properties and failure modes of gyroid structures strongly depend on the relative density. The quasi-static compression tests and simulations exhibited consistent results, indicating that increased relative density enhances the mechanical properties of the structures. Concurrently, Young's modulus and yield stress also exhibit positive power relationships with relative density. Stress and strain concentrations were observed at a 45° angle relative to the loading direction in gyroid structures with varying relative densities. The progressive failures and plateau regions in the stress–strain curves revealed the high deformability of porous gyroid structures. The bending-dominant behaviour and the fractures at a 45° angle during tests suggested shear load as the primary cause of failures.

Adjusting the loading direction influences the strut orientation and load transfer, thereby significantly affecting the mechanical response of gyroid structures. Under compressive load, different loading directions resulted in distinct mechanical properties and failure modes. Structures with struts more closely aligned with the loading direction demonstrated enhanced stiffness. These variations in mechanical properties, dictated by strut orientation, confirmed the anisotropic behaviour of gyroid structures.

In summary, this study contributed to understanding the geometric and mechanical properties of gyroid structures. The correlations and insights provide a robust foundation for the design and application of gyroid structures, particularly for biomedical implants. The ability to precisely tailor gyroid structures, guided by the correlations established in this study, offers the potential to create structures with enhanced functionality and performance.

Supplementary Materials: The following supporting information can be downloaded at: https://www.mdpi.com/article/10.3390/jmmp8060256/s1, Supplementary Data: Geometric correlations of gyroid structures.

Author Contributions: Conceptualization, C.H. and K.D.; methodology, C.H., B.V., and K.D.; formal analysis, C.H., M.G., and D.R.; investigation, C.H., M.G., and D.R.; resources, B.V. and K.D.; data curation, C.H., M.G., and D.R.; writing—original draft preparation, C.H., M.G., and D.R.; writing—review and editing, C.H., B.V., and K.D.; supervision, B.V. and K.D.; funding acquisition, B.V. and K.D. All authors have read and agreed to the published version of the manuscript.

Funding: This research was funded by the CELSA Research Fund and by Internal Funds KU Leuven.

Data Availability Statement: Data available on request due to privacy or ethical restrictions.

Conflicts of Interest: The authors declare no conflicts of interest.

References

1. Kapfer, S.C.; Hyde, S.T.; Mecke, K.; Arns, C.H.; Schröder-Turk, G.E. Minimal Surface Scaffold Designs for Tissue Engineering. *Biomaterials* **2011**, *32*, 6875–6882. [CrossRef] [PubMed]
2. Li, D.; Liao, W.; Dai, N.; Dong, G.; Tang, Y.; Xie, Y.M. Optimal Design and Modeling of Gyroid-Based Functionally Graded Cellular Structures for Additive Manufacturing. *Comput.-Aided Des.* **2018**, *104*, 87–99. [CrossRef]
3. Yuan, L.; Ding, S.; Wen, C. Additive Manufacturing Technology for Porous Metal Implant Applications and Triple Minimal Surface Structures: A Review. *Bioact. Mater.* **2019**, *4*, 56–70. [CrossRef] [PubMed]
4. Kosonen, T.; Kakko, K.; Raitanen, N. Evaluation of Pore Re-Opening after HIP in LPBF Ti–6Al–4V. *Powder Metall.* **2021**, *64*, 425–433. [CrossRef]
5. Liu, J.; Zhang, K.; Liu, J.; Zhang, R.; Zeng, Z.; Zhu, Y.; Huang, A. Discovering the Role of the Defect Morphology and Microstructure on the Deformation Behavior of Additive Manufactured Ti–6Al–4V. *Mater. Sci. Eng. A* **2022**, *859*, 144202. [CrossRef]
6. Cao, S.; Zou, Y.; Lim, C.V.S.; Wu, X. Review of Laser Powder Bed Fusion (LPBF) Fabricated Ti-6Al-4V: Process, Post-Process Treatment, Microstructure, and Property. *Light Adv. Manuf.* **2021**, *2*, 313–332. [CrossRef]
7. Savio, D.; Bagno, A. When the Total Hip Replacement Fails: A Review on the Stress-Shielding Effect. *Processes* **2022**, *10*, 612. [CrossRef]
8. Ataee, A.; Li, Y.; Fraser, D.; Song, G.; Wen, C. Anisotropic Ti-6Al-4V Gyroid Scaffolds Manufactured by Electron Beam Melting (EBM) for Bone Implant Applications. *Mater. Des.* **2018**, *137*, 345–354. [CrossRef]
9. Musthafa, H.-S.N.; Walker, J.; Rahman, T.; Bjørkum, A.; Mustafa, K.; Velauthapillai, D. In-Silico Prediction of Mechanical Behaviour of Uniform Gyroid Scaffolds Affected by Its Design Parameters for Bone Tissue Engineering Applications. *Computation* **2023**, *11*, 181. [CrossRef]
10. Yang, L.; Wu, S.; Yan, C.; Chen, P.; Zhang, L.; Han, C.; Cai, C.; Wen, S.; Zhou, Y.; Shi, Y. Fatigue Properties of Ti-6Al-4V Gyroid Graded Lattice Structures Fabricated by Laser Powder Bed Fusion with Lateral Loading. *Addit. Manuf.* **2021**, *46*, 102214. [CrossRef]
11. Liu, Y.; Rath, B.; Tingart, M.; Eschweiler, J. Role of Implants Surface Modification in Osseointegration: A Systematic Review. *J. Biomed. Mater. Res. A* **2020**, *108*, 470–484. [CrossRef] [PubMed]
12. Hameed, P.; Liu, C.-F.; Ummethala, R.; Singh, N.; Huang, H.-H.; Manivasagam, G.; Prashanth, K.G. Biomorphic Porous Ti6Al4V Gyroid Scaffolds for Bone Implant Applications Fabricated by Selective Laser Melting. *Prog. Addit. Manuf.* **2021**, *6*, 455–469. [CrossRef]
13. Walker, J.M.; Bodamer, E.; Kleinfehn, A.; Luo, Y.; Becker, M.; Dean, D. Design and Mechanical Characterization of Solid and Highly Porous 3D Printed Poly(Propylene Fumarate) Scaffolds. *Prog. Addit. Manuf.* **2017**, *2*, 99–108. [CrossRef]
14. Schoen, A.H. *Infinite Periodic Minimal Surfaces Without Self-Intersections*; NASA Electronics Research Center: Cambridge, MA, USA, 1970.
15. R Core Team. *R: A Language and Environment for Statistical Computing*; R Core Team: Vienna, Austria, 2021.
16. Vrancken, B.; Thijs, L.; Kruth, J.-P.; Van Humbeeck, J. Heat Treatment of Ti6Al4V Produced by Selective Laser Melting: Microstructure and Mechanical Properties. *J. Alloys Compd.* **2012**, *541*, 177–185. [CrossRef]
17. Ashby, M.; Evans, A.; Fleck, N.; Gibson, L.; Hutchinson, J.; Wadley, H.; Delale, F. Metal Foams: A Design Guide. *Appl. Mech. Rev.* **2001**, *54*, B105–B106. [CrossRef]
18. Gibson, L.J. Mechanical Behavior of Metallic Foams. *Annu. Rev. Mater. Sci.* **2000**, *30*, 191–227. [CrossRef]
19. Yang, L.; Yan, C.; Fan, H.; Li, Z.; Cai, C.; Chen, P.; Shi, Y.; Yang, S. Investigation on the Orientation Dependence of Elastic Response in Gyroid Cellular Structures. *J. Mech. Behav. Biomed. Mater.* **2019**, *90*, 73–85. [CrossRef]
20. Yan, C.; Hao, L.; Yang, L.; Hussein, A.Y.; Young, P.G.; Li, Z.; Li, Y. Chapter 6—Numerical Analysis on the Mechanical Properties of Uniform TPMS. In *Triply Periodic Minimal Surface Lattices Additively Manufactured by Selective Laser Melting*; Yan, C., Hao, L., Yang, L., Hussein, A.Y., Young, P.G., Li, Z., Li, Y., Eds.; 3D Printing Technology Series; Academic Press: Cambridge, MA, USA, 2021; pp. 183–217. ISBN 978-0-12-824438-8.
21. Song, C.; Chen, J.; Lei, H.; Yang, Z.; Deng, Z.; Li, Y.; Wang, J.; Yang, Y.; Han, C. Radial Gradient Design Enabling Additively Manufactured Low-Modulus Gyroid Tantalum Structures. *Int. J. Mech. Sci.* **2024**, *262*, 108710. [CrossRef]
22. Torres-Sanchez, C.; Al Mushref, F.R.A.; Norrito, M.; Yendall, K.; Liu, Y.; Conway, P.P. The Effect of Pore Size and Porosity on Mechanical Properties and Biological Response of Porous Titanium Scaffolds. *Mater. Sci. Eng. C* **2017**, *77*, 219–228. [CrossRef]
23. Parisien, A.; ElSayed, M.S.A.; Frei, H. Mechanoregulation Modelling of Stretching versus Bending Dominated Periodic Cellular Solids. *Mater. Today Commun.* **2022**, *33*, 104315. [CrossRef]
24. Parisien, A.; ElSayed, M.S.A.; Frei, H. Mature Bone Mechanoregulation Modelling for the Characterization of the Osseointegration Performance of Periodic Cellular Solids. *Materialia* **2022**, *25*, 101552. [CrossRef]
25. Li, L.; Shi, J.; Zhang, K.; Yang, L.; Yu, F.; Zhu, L.; Liang, H.; Wang, X.; Jiang, Q. Early Osteointegration Evaluation of Porous Ti6Al4V Scaffolds Designed Based on Triply Periodic Minimal Surface Models. *J. Orthop. Transl.* **2019**, *19*, 94–105. [CrossRef] [PubMed]

26. Yao, Y.; Qin, W.; Xing, B.; Sha, N.; Jiao, T.; Zhao, Z. High Performance Hydroxyapatite Ceramics and a Triply Periodic Minimum Surface Structure Fabricated by Digital Light Processing 3D Printing. *J. Adv. Ceram.* **2021**, *10*, 39–48. [CrossRef]
27. Barba, D.; Alabort, E.; Reed, R.C. Synthetic Bone: Design by Additive Manufacturing. *Acta Biomater.* **2019**, *97*, 637–656. [CrossRef]
28. Yang, L.; Yan, C.; Cao, W.; Liu, Z.; Song, B.; Wen, S.; Zhang, C.; Shi, Y.; Yang, S. Compression–Compression Fatigue Behaviour of Gyroid-Type Triply Periodic Minimal Surface Porous Structures Fabricated by Selective Laser Melting. *Acta Mater.* **2019**, *181*, 49–66. [CrossRef]
29. Fan, Z.; Huang, G.; Lu, Y.; Chen, Y.; Zeng, F.; Lin, J. Full Compression Response of FG-Based Scaffolds with Varying Porosity via an Effective Numerical Scheme. *Int. J. Mech. Sci.* **2022**, *223*, 107294. [CrossRef]
30. Cutolo, A.; Engelen, B.; Desmet, W.; Van Hooreweder, B. Mechanical Properties of Diamond Lattice Ti–6Al–4V Structures Produced by Laser Powder Bed Fusion: On the Effect of the Load Direction. *J. Mech. Behav. Biomed. Mater.* **2020**, *104*, 103656. [CrossRef]
31. Morgan, E.F.; Unnikrisnan, G.U.; Hussein, A.I. Bone Mechanical Properties in Healthy and Diseased States. *Annu. Rev. Biomed. Eng.* **2018**, *20*, 119–143. [CrossRef]
32. Liu, B.; Wang, H.; Zhang, N.; Zhang, M.; Cheng, C.-K. Femoral Stems With Porous Lattice Structures: A Review. *Front. Bioeng. Biotechnol.* **2021**, *9*, 772539. [CrossRef]
33. Han, C.; Wang, Y.; Wang, Z.; Dong, Z.; Li, K.; Song, C.; Cai, C.; Yan, X.; Yang, Y.; Wang, D. Enhancing Mechanical Properties of Additively Manufactured Voronoi-Based Architected Metamaterials via a Lattice-Inspired Design Strategy. *Int. J. Mach. Tools Manuf.* **2024**, *202*, 104199. [CrossRef]
34. Challis, V.J.; Xu, X.; Zhang, L.C.; Roberts, A.P.; Grotowski, J.F.; Sercombe, T.B. High Specific Strength and Stiffness Structures Produced Using Selective Laser Melting. *Mater. Des.* **2014**, *63*, 783–788. [CrossRef]
35. Cuadrado, A.; Yánez, A.; Martel, O.; Deviaene, S.; Monopoli, D. Influence of Load Orientation and of Types of Loads on the Mechanical Properties of Porous Ti6Al4V Biomaterials. *Mater. Des.* **2017**, *135*, 309–318. [CrossRef]
36. Wu, Z.; Narra, S.P.; Rollett, A. Exploring the Fabrication Limits of Thin-Wall Structures in a Laser Powder Bed Fusion Process. *Int. J. Adv. Manuf. Technol.* **2020**, *110*, 191–207. [CrossRef]
37. Kranz, J.; Herzog, D.; Emmelmann, C. Design Guidelines for Laser Additive Manufacturing of Lightweight Structures in TiAl6V4. *J. Laser Appl.* **2014**, *27*, S14001. [CrossRef]
38. Shange, M.; Yadroitsava, I.; du Plessis, A.; Yadroitsev, I. Roughness and Near-Surface Porosity of Unsupported Overhangs Produced by High-Speed Laser Powder Bed Fusion. *3D Print. Addit. Manuf.* **2022**, *9*, 288–300. [CrossRef]
39. Hou, C.; Sinico, M.; Vrancken, B.; Denis, K. Investigation of the Laser Powder Bed Fusion Manufacturing Process and Quasi-Static Behaviour of Ti6Al4V Voronoi Structures. *J. Mater. Process. Technol.* **2024**, *328*, 118410. [CrossRef]